T5-CQF-489

3 3073 00030678 5

Regulation and Functional Significance of T-Cell Subsets

Chemical Immunology

Vol. 54

Series Editors
Frank W. Fitch, Chicago, Ill.
Kimishige Ishizaka, La Jolla, Calif.
Peter J. Lachmann, Cambridge
Byron H. Waksman, New York, N.Y.

KARGER

Basel · Freiburg · Paris · London · New York · New Delhi · Bangkok · Singapore · Tokyo · Sydney

Regulation and Functional Significance of T-Cell Subsets

Volume Editor
Robert L. Coffman, Palo Alto, Calif.

30 figures and 10 tables, 1992

KARGER

Basel · Freiburg · Paris · London · NewYork · New Delhi · Bangkok · Singapore · Tokyo · Sydney

SETON HALL UNIVERSITY
McLAUGHLIN LIBRARY
SO. ORANGE, N. J.

Chemical Immunology

Formerly published as 'Progress in Allergy'
Founded 1939 by Paul Kallòs

RC
583
.P7
v. 54
1992

Bibliographic Indices
   This publication is listed in bibliographic services, including Current Contents® and Index Medicus.

Drug Dosage
   The authors and the publisher have exerted every effort to ensure that drug selection and dosage set forth in this text are in accord with current recommendations and practice at the time of publication. However, in view of ongoing research, changes in government regulations, and the constant flow of information relating to drug therapy and drug reactions, the reader is urged to check the package insert for each drug for any change in indications and dosage and for added warnings and precautions. This is particularly important when the recommended agent is a new and/or infrequently employed drug.

All rights reserved.
   No part of this publication may be translated into other languages, reproduced or utilized in any form or by any means, electronic or mechanical, including photocopying, recording, micro-copying, or by any information storage and retrieval system, without permission in writing from the publisher.

©   Copyright 1992 by S. Karger AG, P.O. Box, CH–4009 Basel (Switzerland)
   Printed on acid-free paper.
   ISBN 3-8055-5577-6

Contents

Contents

Contents

Murine CD4⁺ T- Cell Subsets Generated by Antigen-Independent and Dependent Mechanisms

Induction, Regulation and Function of T-Cell Subsets in Leishmaniasis

Contents

Contents   IX

Coffman RL (ed): Regulation and Functional Significance of T-Cell Subsets.
Chem Immunol. Basel, Karger, 1992, vol 54, pp 1–20

Programming of Lymphocyte Responses to Activation: Extrinsic Factors, Provided Microenvironmentally, Confer Flexibility and Compartmentalization to T-Cell Function[1]

Raymond A. Daynes, Barbara A. Araneo

Department of Pathology, University of Utah School of Medicine,
Salt Lake City, Utah, USA

Introduction

The most important function of the immune system is to provide its host with protection against diseases. To carry out these tasks, a large and diverse array of effector mechanisms have evolved, the majority of which exhibit antigen specificity. Each individual effector mechanism possesses a degree of uniqueness with respect to its ability to influence the rate of progression, to detoxify, or to promote the elimination of microbial pathogens or tumor cells. Such a diversity in available mechanisms is absolutely essential since no single effector response can effectively deal with all forms of pathogenic insults. Furthermore, to protect normal function of the various nonlymphoid organ systems and tissues of the body, requires careful selection, activation, and compartmentalization of the most appropriate types of immune effector mechanisms. Equally important is the simultaneous capacity to down-regulate the development of other types of responses. Immunologic effector responses must therefore be both effective and practical, and at the same time be appropriately regulated anatomically to reduce the risk of pathologic consequences.

Understanding the mechanisms by which the immune system selects and promotes the development of particular responses following an antigenic insult, represents a central issue in immunobiology. Appropriate discrimina-

[1] Supported in part by US Public Health Service Grants No. CA25917 and CA33065, awarded by the National Institutes of Health.

tion requires consideration of the agents providing the antigenic insult as well as the inputs being simultaneously supplied by the organs and tissues being affected. Only after all of this essential information is received can the immune system perform in the most appropriate manner. In the case of a tissue infection, the pathogenic microorganisms impart specificity to the system in the way of foreign antigen. It is probable that other consequences of infection, either directly or indirectly, provide necessary input signals that aid the immune system in the selection of appropriate modes of responsiveness.

The nonlymphoid tissues and organs of the body, which work collectively to sustain the life of the host, must also be capable of providing regulatory information to cells of the immune system. This information, mediated through the activities of inflammation-induced tissue cytokines, prostaglandins, plus other types of biological response modifiers, becomes integrated into the complex equation to control the mechanisms which regulate effector response selection.

T cells, through their capacity to produce a number of lymphokines in response to activation, play a central role in guiding the development of immune effector responses. Mechanisms which operate to control the synthesis and secretion of these pleiotropic biologic response modifiers therefore, directly influence the quantitative and qualitative nature of immunity. The lymphokines and cytokines provide important information, not only to cells of the immune system, but also to cells of the other tissue and organ systems. For this information to be meaningful, it is essential that lymphokine production remains tightly controlled at the levels of both cellular source and duration. Autocrine and paracrine effects by lymphokines and cytokines should be the norm, since only a few species are capable of working effectively when provided via endocrine routes. These essential anatomic restrictions, therefore, cannot be adequately provided by bolus injection of recombinant materials, and may explain the limited success associated with this form of therapy.

The vast majority of the T cells in the peripheral circulation are known to reside within the recirculating T-cell pool. These cells continuously enter and exit secondary lymphoid organs throughout the body, maintaining residence within any particular site for only finite periods of time. Over the lifespan of any individual mature T cell, therefore, it has probably taken up temporary residence in most of a host's secondary lymphoid organs. T-cell recirculation provides the immune system with a means for clonally-restricted T cells to provide a level of surveillance over all the tissue and organ systems. It is important to know whether the functional capabilities of a T cell are preset, or whether its potential for involvement in immunologic responses exhibits flexibility.

Are T Cells Genetically Programmed for a Restriction in Both Their Specificity and Function?

It is universally accepted that most T cells acquire their specificity for antigen, and a self-MHC-restricting element, during processes which occur during their ontogeny within the thymus. Less well understood, however, is the extent to which intrathymic maturation confers genetic restrictions upon individual T cells that regulate their potential for immunologic involvement.

Based upon clear distinctions in the patterns of lymphokines capable of being synthesized and secreted by panels of cloned T-cell lines, a prominent theory in cellular immunology presently advocates the existence, in vivo, of multiple T-cell subsets which differ from one another in their functional potential [1–4]. This model implies that each of these distinct subsets of T cells expresses the total T-cell receptor repertoire. How the immune system, in any given circumstance, selects the T-cell subset capable of producing the pattern of lymphokines most appropriate for driving the development of a given response is presently unknown.

Investigators have repeatedly observed that the activation of T cells from immunologically naive donors in vitro results in the production of large amounts of IL-2 and minimal synthesis of other T-cell cytokines [5–8]. At later time periods following activation, T cells from these same cultures acquire the capacity to produce multiple lymphokines, suggesting their differentiation from immature, naive T cells to a more mature phenotype [7, 8]. Results from these types of experiments have been used to support a model which contains the belief that, in vivo, naive T cells undergo differentiation events which unidirectionally move them down a pathway from being IL-2 producers (T_HP) through the stage where they possess a very broad potential to produce lymphokines (T_H0) and ultimately into a fully differentiated phenotype where their capacity to produce lymphokines becomes quite restricted (T_H1 or T_H2) [7].

There exists an extensive amount of recent evidence which demonstrates that some of the lymphokines themselves, as a reflection of one of their pleiotropic activities, can control the synthesis or the biologic function of other lymphokines [3, 9–12]. It has been concluded that these regulatory functions by particular lymphokine species play essential roles in maintaining the synthesis of a particular pattern of lymphokines by T cells responding to antigen [13, 14]. Logically, however, lymphokine control of lymphokine synthesis and/or function must represent a downstream regulatory event in the genesis of immunologic responses. Furthermore, minimal information has emerged from these models concerning the mechanisms which operate to control the initial selection for activation of T-cell types which produce appropriate (as defined by the particular system) lymphokine patterns.

It could be easily argued that the immune system requires mechanisms that function to carefully regulate the pattern of lymphokines produced by T cells following their activation. It is also quite probable that the patterns of lymphokines produced by T cells in response to antigenic stimulation serve to successfully guide the development of particular effector responses. What is questionable, however, is whether the immune system requires the existence of precommitted cells to facilitate such a diverse range of functions. Alternatively, it might be argued that every T cell contains the genes and the potential to produce all of the T-cell lymphokines, but the nature of its response to activation is under microenvironmental control. Such a model would predict that a T-cell residing within a given lymphoid organ at the time of its activation, might be stimulated to produce a particular pattern of lymphokines including those lymphokines which could cross-regulate the function/synthesis of other species [12–14]. This exact same T cell might produce a totally different pattern of lymphokines if it was activated by antigen while residing within some other lymphoid organ or tissue site.

The primary objective of this paper is to present results from some of our recent studies that collectively support an alternative model concerning the mechanisms which operate in vivo to control T-cell involvement in immunologic responses. Instead of a requirement for multiple T-cell subsets having overlapping antigen specificities and MHC restriction elements, our conservative model suggests that a T cell's involvement in an immunologic response is under environmental control. The substances which guide the responses elicited by T cells to antigen are under the quantitative and qualitative regulation by factors generally considered extrinsic to the immune system. Since the biosynthesis of these factors is microenvironmentally regulated, their influences on the functional capabilities of T cells become anatomically compartmentalized.

Steroid Hormone Control Over the Qualitative Nature of
T-Cell Responses

In 1989 [15], we presented evidence which indicated that T cells are affected in a consistent manner following their exposure to the modulatory influences of glucocorticoids (GCS). Regardless of whether we employed heterogeneous populations of T cells from naive donors, antigen-primed T cells, cloned T-cell lines, or T-cell hybridomas, they all demonstrated consistency with regards to the changes caused by GCS exposure. Following a direct treatment of T cells with physiologic levels of GCS, their capacity to produce the lymphokine IL-4 in response to activation was found to be greatly enhanced. This elevated potential to produce IL-4 was evident following exposure to GCS levels between 1 and 10% of those required to inhibit the production of either IL-2 or γIFN. Further, the ability of GCS to enhance the ability of T cells to produce IL-4 required their activation under serum-free conditions.

Some very important differences existed between our finding and those reported by others [5–8]. Results from all of our experiments consistently demonstrated that T cells were capable of simultaneously producing multiple lymphokines following activation. Although some investigators have reported findings similar to ours [16–18], many investigators have observed that naive T cells are restricted to produce primarily IL-2 following their activation and that only 'memory cells' gain the capacity to produce other lymphokine species [6–8, 19]. We now understand the mechanism responsible for this apparent discrepancy and its relationship to the presence of serum supplements, and will provide a detailed explanation in a future section of this article.

Finding that T cells exposed to GCS in vitro or in vivo produced a different pattern of lymphokines from normal control T cells, provided a plausible explanation to our previous demonstration that T cells, isolated from animals stressed by high-dose exposure to ultraviolet radiation (UVR), produce low levels of IL-2 and γIFN and enhanced amounts of IL-4 in response to activation [20]. It is now widely accepted that one major consequence of UVR exposure is the stimulation of keratinocyte production of IL-1 and other potent inflammatory cytokines [21]. These cytokines, through their ability to directly stimulate the hypothalamic-pituitary-adrenal axis, stimulate release of GCS [22].

We attempted to establish a cause-effect relationship between fluctuations in endogenous GCS levels, and the changes in lymphokine patterns by

T cells obtained from UVR-exposed donors. In these studies, adrenal output of GCS was temporarily blocked by the systemic administration of the drug metyrapone, a potent 11β-hydroxylase inhibitor. Following the exposure of these animals to UVR, their T cells were collected, stimulated, and assayed for the production of lymphokines. We found that T cells from metyrapone-treated, UVR-exposed donors retained a normal capacity to produce IL-2, suggesting the causal involvement of GCS in the immunomodulatory influences of UVR (fig. 1).

Our experiments conducted on animals being treated with metyrapone, provided us with an additional reward. As presented in figure 1, we consistently observed that the T cells from either lymph nodes or spleens of these animals, produced greatly elevated levels of IL-2 in response to activation when compared to the lymphokines produced by cells taken from syngeneic controls. By gaining a better appreciation of the changes in steroid hormone synthesis caused by metyrapone treatment, we first became introduced to the steroid hormone, dehydroepiandrosterone (DHEA).

Metyrapone inhibition of GCS production in vivo results in an increased release of pituitary ACTH. The elevations in ACTH are due to the elimination of the feedback control over its production, an effect which is normally mediated by GCS. With GCS production being inhibited, ACTH then causes an enhanced rate of production of the weak adrenal androgen DHEA. Enhanced levels of DHEA are well recognized to represent a normal consequence of metyrapone inhibition of endogenous GCS production.

DHEA Regulation of T Cells

DHEA represents a steroid hormone that has been extensively studied for many years. It has been previously reported to be involved in a wide variety of physiologic, immunologic, and pathologic conditions [for reviews, see 23, 24]. Most endocrinologists believe that the primary function of DHEA is to serve as a precursor for the synthesis of testosterone and the estrogens by the gonads. Our recent finding that DHEA can directly enhance the ability of activated T cells to produce IL-2 (and γIFN) [25], and that DHEA-responsive T cells possess high-affinity intracellular DHEA receptors [26], strongly suggests that this steroid hormone may also have profound influences on the immune system.

DHEA possesses the interesting characteristic of being end-organ metabolized from a circulating precursor form. Prior to its release into the

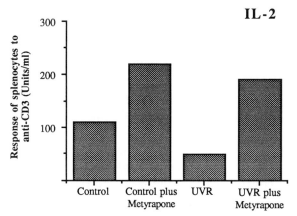

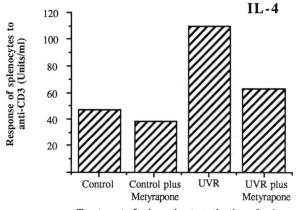

Fig. 1. Treatment of C3H/HeN mice with metyrapone, an 11β-hydroxylase inhibitor, abrogates the ability of UVR exposure to depress T-cell production of IL-2. Biodegradable pellets containing 12.5 mg metyrapone (Innovative Research of America, Toledo, Ohio) were subcutaneously implanted into a group of shaved normal mice. Half of these animals, plus an equal number of untreated normal controls, were then exposed to UVR (15 kJ/m²/mouse of UVB from FS-40 fluorescent bulbs). Three days later, spleens were removed from animals within the experimental and control groups, dissociated into single-cell suspensions, and activated with 1 μg/ml of anti-CD3ε under serum-free conditions. IL-2 and IL-4 levels in the cell supernatants after 24 h were quantitated employing the HT-2 bioassay as previously described [20].

bloodstream, the vast majority of newly synthesized DHEA becomes sulfated. This sulfated form of DHEA (DHEAS) represents the major steroid hormone in the circulation of humans, and is present at levels between 1 and 6 μg/ml in normal adults. It is also known that very young and very old humans produce far less DHEA and DHEAS than individuals between 6 and 40 years of age. We found that as DHEAS, the steroid hormone was totally incapable of directly affecting the functional properties of T cells [24]. Further, we established that a direct correlation existed between the DHEA sulfatase content (DHEAS → DHEA) within a particular lymphoid organ, and the capacity of T cells isolated from that lymphoid organ to produce IL-2 and γIFN following activation [27]. Interestingly, the highest DHEA sulfatase levels, and therefore the greatest T-cell potential for IL-2 production, were found in those lymphoid organs which receive their primary afferent lymphatic drainage from nonmucosal tissues (e.g., skin). High DHEA sulfatase activity was also found in the spleen. The lowest DHEA sulfatase levels were found in the Peyer's patches, and the deep cervical, periaortic, and parathymic lymph nodes. T cells residing in these mucosal tissue draining lymphoid organs were found to produce high levels of IL-4 (and IL-5) and low levels of IL-2 (and γIFN) following their stimulation in vitro [27]. Again, it must be emphasized that these functional distinctions could only be made with lymphocytes maintained under serum-free conditions in vitro.

Results from these studies also determined that antigen-primed T cells, isolated from lymphoid organs receiving their predominant drainage from either mucosal or nonmucosal tissues, would, after adoptive transfer to normal syngeneic recipients, respond to antigen with lymphokine patterns controlled by new lymphoid organ environment. These results strongly suggested that a T cell's response to antigen possesses flexibility, and is under the regulation by extrinsic factors present within the immediate cellular microenvironment.

Additional Steroid Hormones Involved in T-Cell Regulation

DHEA represents a steroid hormone that appears to require end-organ metabolism from its circulating precursor form (DHEAS). A number of other steroid hormones exist which share this important property. These include dihydrotestosterone (DHT), which is metabolized by 5α-reductase from its circulating precursor testosterone, and 1,25-dihydroxyvitamin D_3 (1,25(OH)$_2$D$_3$) which is metabolized by 1α-hydroxylase from 25-hydroxyvitamin D$_3$. We have recently established, once again by employing serum-free conditions, that both of these steroid hormones can profoundly influence the functional potential of T cells. DHT selectively reduces the ability of T cells to produce IL-4, IL-5, and γIFN following activation, without affecting their capacity to produce IL-2 [28]. 1,25(OH)$_2$D$_3$ exerts quite a different effect on T-cell function. In addition to its already appreciated depressive effects on the synthesis of IL-2 and γIFN [29, 30], 1,25(OH)$_2$D$_3$ at much lower doses can significantly enhance the ability of activated T cells to produce IL-4 and IL-5 [29, 30]. The circulating precursors of these two steroid hormones were established to have little effect on the functional properties of T cells.

Predictions Concerning Steroid Hormone Control over T-Cell Function

Collectively, our findings support the general concept that the genetic programs of resting recirculating T cells are continuously being altered by extrinsic environmental influences. We believe that steroid hormones, either presented in their active forms systemically (e.g. GCS), or being provided to T cells only within discrete microenvironments as a consequence of end-organ metabolism (e.g., DHEA, DHT, or 1,25(OH)$_2$D$_3$), perform important roles in this process. The steroid hormones represent very logical candidates for this type of complex control, since individual species are capable of activating potent hormone-specific transcriptional factors. These activated transcriptional factors are then able to regulate the activities of a large number of cellular genes, thereby quantitatively and qualitatively controlling the potential of a T cell's response to activation. Our model implies, therefore, that basal regulation of the immune system at the level of the T cell requires the continual presence of the needed substrates (prohormones). The anatomic compartmentalization of functional potential for T cells, therefore, would be dependent on the cellular source of the steroid hormone metabolizing enzymes able to convert the steroid hormone substrates to their bioactive species. We now appreciate that macrophages can contain each of these enzymes [28–30].

Consequences of Dysregulation of Steroid Hormone Metabolism on the Immune System and Immunologic Function

There exist numerous examples that could be used to strengthen the important relationship that must exist between the endocrine and immune systems. It is possible, for example, that some of the immunologic changes which occur with the disease sarcoidosis may be derived from the 1α-hydroxylase abnormalities that exist in alveolar macrophages from these patients [31]. Alterations in the regulation of 25-hydroxyvitamin D_3 metabolism, allowing $1,25(OH)_2D_3$ to become an endocrine acting hormone, could be responsible for the immunologic changes observed in active disease [32]. Another example might involve individuals exhibiting hyperresponsiveness to stress resulting in high GCS levels. Such individuals might be rendered more susceptible to certain infectious diseases [33]. Conversely, low-stress responders may be more susceptible to some types of autoimmune diseases [33].

The best example to make the point concerning the consequences of changes to steroid hormones, however, is associated with the natural process of aging. The immune system becomes far less efficient with advancing age [34, 35]. A tremendous number of age-associated changes to selected components of the immune system have been reported to take place, making it quite unattractive to get old [35]. The terms 'defects' and 'abnormalities' are in general usage throughout this literature, implying that the diminished immunocompetence of the elderly results from intrinsic changes having pathologic consequences. We would like to present an alternative viewpoint, one which represents almost the antithesis to some of the present paradigms concerning the mechanisms responsible for immunosenescence. Might it be possible that the immune system of elderly individuals is functioning in a completely normal manner? That is to say, are the cells of the elderly, plus all of the molecular and biochemical mechanisms which collectively comprise their immune systems, capable of responding to exogenous and endogenous perturbations exactly as instructed? Assuming this is the case, and realizing that the consequences which follow an antigenic insult to the elderly often fails to meet its needs, leads to the possibility that changes to essential extrinsic regulatory processes may be involved. Some of these should be easily correctable.

As stated previously, endogenous DHEA production in vivo exhibits age dependence. Biochemically, this occurs through an age-related reduction in the ability of the enzyme $P450_{17\alpha}$ to efficiently convert 17α-hydroxypregnenolone to DHEA. As a consequence, circulating DHEAS levels in the plasma of all mammals goes down dramatically with advancing age. From our studies involving the direct effects of DHEA on T-cell function [25], plus the results of additional experiments which established that in vivo, DHEA influences to T cells were probably compartmentalized to particular lymphoid organs

[27], we reasoned that providing supplemental DHEAS to old animals might be able to reverse some of the most evident age-related alterations to the immune system [36]. Our initial experiments in this area have involved providing DHEAS supplementation to normal mice greater than 24 months of age. T cells were isolated from various lymphoid organs of these animals and compared to identical lymphoid organ sources of T cells from mature adult and normal old (24+ months) donors. The consequences of aging were found to result in profound depressions in the amounts of IL-2, IL-3, and GM-CSF capable of being synthesized, with simultaneous increases in the amount of IL-4, IL-5, and γIFN being observed (fig. 2). Of great interest was the finding that T cells from old animals on DHEAS supplementation responded almost identically to T cells from mature adult donors. Therefore, it appears that many of the consistently observed alterations in T-cell lymphokine patterns which accompany age, may actually be secondary manifestations of a primary loss in circulating DHEAS. This suggestion was further strengthened by the finding that old animals provided with an exogenous source of DHEAS are fully capable of mounting normal immunologic responses following antigen administration (fig. 3). This type of experiment has been repeated numerous times, employing a variety of test antigens, with the results consistently demonstrating normal immune responses being elicited by the old animals provided with DHEAS supplementation [unpubl. data].

There are many clinical conditions, including aging, stress, trauma, pregnancy, neoplasia, and autoimmunity, where the immune system can be compromised by mechanisms involving alterations to the function, and not the physical presence, of its required cellular elements. Many of these functional alterations may actually be caused by fluctuations in the availability of extrinsic substances that are essential to normal immunoregulation. We have focused our attention on one such group of substances, the steroid hormones, although other factors may be equally important to normal immunologic homeostasis. These include the roles played by prostaglandins [37], growth hormones [38], prolactin [39], α_2-macroglobulin [40], and even secreted soluble forms of some of the cytokine and lymphokine receptors [41].

Platelet-Derived Growth Factor as a Regulator of T-Cell Function

Over the entire time period of our studies involving the analysis of potential immunoregulatory roles played by steroid hormones, we have

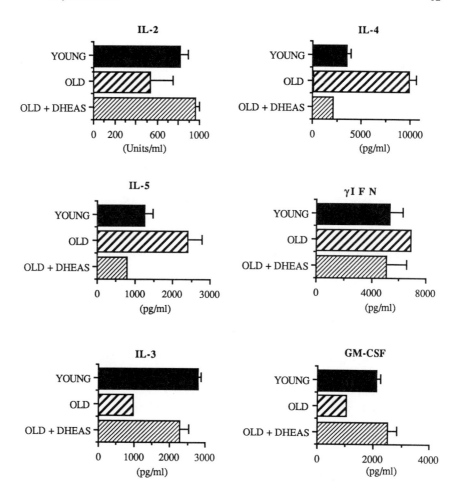

Fig. 2. Old C3H/HeN strain (> 2 years) provided with supplemental DHEAS (100 µg/ml in drinking water) fail to exhibit age-related changes in the ability of their splenocytes to produce lymphokines and cytokines following activation. The spleen's peripheral and mucosal lymph nodes were aseptically removed from groups of young (15-week), old, and age-matched old animals which were maintained on supplemental DHEAS (started when the animals were 1 year of age). Lymphocytes from individual mice were cultured serum-free in the presence or absence of 1 µg/ml anti-CD3ε. After 24 h, the supernatants were collected and the lymphokine and cytokine content quantitatively analyzed. IL-2 levels were analyzed by bioassay. All other cytokines were analyzed by capture ELISA employing reagents obtained from PharMingen (San Diego, Calif.). The results presented are those obtained from the splenocyte cultures. Although the patterns of lymphokines produced following activation exhibit lymphoid organ specificity, all age-related changes were eliminated through supplemental DHEAS treatment [36].

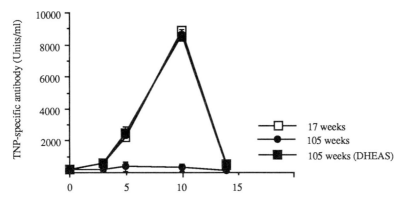

Fig. 3. Old mice (> 2 years) provided with supplemental DHEAS (100 µg/ml in drinking water) respond normally to immunization with TNP-modified horse erythrocytes (HRBC). Parallel groups of young, old, and old C3H/HeN strain mice provided with supplemental DHEAS, were subcutaneously immunized with TNP-HRBC (25 µl of a 20% suspension). Three, 5, 10, and 14 days after immunization, serum samples were collected and individually analyzed for anti-TNP antibody titers by ELISA. The results (± SD) are presented in units/ml where 1 unit is equivalent to 1/dilution where the response was half saturating. This experiment has been conducted numerous times with various hapten and protein antigens with equivalent results.

remained concerned about the differences between our findings and those being reported by others [5–8, 19]. Most disturbing was our inability to observe that T cells, freshly isolated from normal naive animals, were restricted in the patterns of lymphokines produced following activation [15, 25, 42]. Teleologically, such a restriction represented a logical possibility, as the successful development of immunity must involve the expansion of clonally restricted T cells. Mechanisms to restrict T cells to the predominant production of IL-2, an essential and potent growth factor, seemed like a logical way to satisfy such a need.

We resolved this seemingly important issue by conducting experiments that allowed a comparison between the differences in experimental protocols which existed between our studies. Due to our interest in the effects of steroid hormones, we chose at the beginning to conduct all of our in vitro studies under serum-free conditions. This avoided any possible input by the steroid hormones present in all serum supplements. In retrospect, this choice has proven to be correct, but for reasons other than involving unknown levels of contamination by steroid hormones.

We experimentally questioned the pattern and levels of lymphokines produced by activated T cells, either heterogeneous populations or T-cell hybridomas adapted to serum-free growth conditions. These cells are activated under either serum-free or serum-containing conditions. As previously demonstrated by us, T cells in serum-free medium were able to simultaneously produce multiple lymphokine species following activation [15, 25, 30, 36]. It was also observed that in the presence of serum, these same T cells were highly restricted to the production of only IL-2, and this lymphokine was produced at well above normal levels [42]. The ability of these activated cells to produce IL-4, IL-5, and γIFN was found to be markedly depressed. We reasoned that something in serum must be responsible for the observed restriction in T-cell function.

Serum contains a number of components not found in plasma, many of which are able to exert very profound influences on cellular behavior. Most notable are the growth factors that are normally sequestered within platelets, but are released as a consequence of blood clotting into the serum. The presence of these growth factors is essential for serum to serve as a growth supplement for most non-transformed cell types.

Lymphocytes, placed into culture containing a source of serum, generally fail to proliferate or secrete lymphokines unless activated. This has led to the assumption (unfortunately) that no components present within the culture system itself are exerting influences on the function or potential function of the cells. This has represented a tremendous leap of faith, taken by almost everyone. Some of the consequences of making this assumption became evident when we activated either heterogeneous or homogeneous populations of T cells in medium containing nanogram quantities of platelet-derived growth factor (PDGF). These studies established that PDGF, a normal component of *all* serum sources, limited the T cells to primarily IL-2 production, and markedly inhibited their ability to secrete other lymphokines [42]. Only at later time points, when the T cells actually became

---

Fig. 4. PDGF (1 ng/ml of recombinant human PDGF-BB) alters the capacity of both CD4+ (Hd13.2) and CD8+ (DB8-30.7) T-cell hybridomas to produce lymphokines in response to activation with immobilized anti-CD3ε. T-cell hybridomas, adapted to propagate under serum-free conditions, were added to tissue culture plates containing immobilized anti-CD3ε. Supernatants were collected following a 24-hour incubation, and assayed for lymphokine content by bioassay (IL-2) or by capture ELISA (IL-4 and IL-5). Similar results are obtained from experiments using freshly isolated lymph node or spleen lymphocytes obtained from normal murine donors [42].

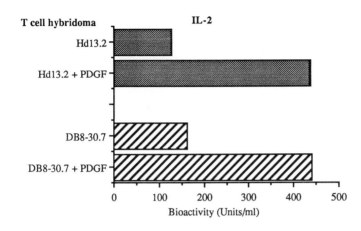

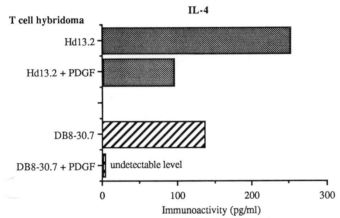

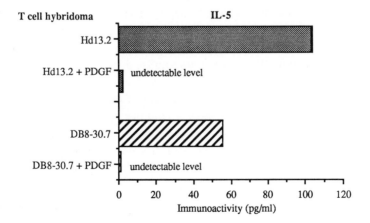

'desensitized' to PDGF-mediated effects, were the cells able to produce the lymphokines which are fully capable of immediate synthesis by control T cells (namely IL-4, IL-5, and γIFN). An example of the results from this type of experiment is presented in figure 4. These findings may necessitate a reappraisal of the data supporting the concept that peripheral T cells somehow mature functionally following cellular activation [5–8, 19].

We feel strongly that the observed influences of PDGF on T-cell function are more than simple artifacts caused by in vitro culture in serum-containing medium. In addition to preformed PDGF being present in platelets, this growth factor can also be produced by macrophages and some other cell types following appropriate stimulation [43, 44]. Additional sophistication to this lymphocyte-regulatory system emanates from some very recent work in the PDGF field [45, 46]. The existence of various isoforms of PDGF dimers, their unique receptor-binding characteristics, and the distinct types of membrane receptors that exist to control PDGF-mediated responses, indicate that the activities of this growth factor are probably intimately integrated into the regulation of lymphocyte function.

Overall Conclusions

Parallel with the recent integration of molecular approaches into the field of immunology which have provided investigators with a whole set of new research tools, has come the increasing practice of asking highly focused and sophisticated questions. While this tactic has proven quite useful in some systems, it is almost certain to create new sets of problems. The most obvious problem is associated with the ongoing belief that full understanding of the immune system can only be accomplished when it is experimentally evaluated under conditions which physically separate it from any influences by other organ systems. While this reductionist approach is essential for advancing some scientific areas (e.g., regulation of the IL-2 gene by transcription factors which bind within 300 bp upstream of the promoter region), it may fail totally in the analysis of how a complex organ system, like the immune system, operates effectively in vivo.

It has become quite clear to us that the various cell types which classically comprise the immune system (T cells, B cells, macrophages, etc.) can only carry out their functions effectively, by maintaining a constant communication with all the tissue and organ systems. Only then can these cell types acquire necessary information about their position within the body, the tissues which require their protection, and the nature of any antigenic insult. These inputs are absolutely essential, as the types of effector responses elicited can be just as important as generating a response itself. It is easy to envision that influences provided by the steroid hormones, so essential to the homeostasis and development of most organ systems, are also important to the immune system. Equally logical is the acceptance of a role by PDGF, and other growth factors present within platelets, in the regulation of T-cell function in vivo.

References

1 Mosmann TR, Cherwinski H, Bond MW, Gledlin MA, Coffman RL: Two types of murine helper T-cell clone. I. Definition according to profiles of lymphokine activities and secreted proteins. J Immunol 1986;136:2348–2357.
2 Cherwinski HM, Schumacher JH, Brown KD, Mosmann T: Two types of helper T-cell clone. III. Further differences in lymphokine synthesis between TH1 and TH2 clones revealed by RNA hybridization, functionally monospecific bioassays, and monoclonal antibodies. J Exp Med 1987;166:1229–1244.
3 Florentino DF, Bond MW, Mosmann TR: Two types of mouse T helper cell. IV. TH2 clones secrete a factor that inhibits cytokine production by TH1 clones. J Exp Med 1989;170:2081–2095.
4 Street NE, Schumacher JH, Fong TAT, Bass H, Fiorentino DF, Leverah JA, Mosmann TR: Heterogeneity of mouse helper T cells: evidence from bulk cultures and limiting dilution cloning for precursors of TH1 and TH2 clones. J Immunol 1990;144:1629–1639.
5 Salmon M, Kitas GD, Bacon PA: Production of lymphokine mRNA by CD45R$^+$ and CD45R$^-$ helper T cells from human peripheral blood and by human CD4$^+$ T cell clones. J Immunol 1989;143:907–912.
6 Ben-Sasson SZ, LeGros G, Conrad DH, Finkelman FD, Paul WE: IL-4 production by T cells from naive donors: IL-2 required for IL-4 production. J Immunol 1990;145:1127–1136.
7 Swain SL, McKenzie DT, Weinberg AD, Hancock W: Characterization of T helper 1 and 2 cell subsets in normal mice. Helper T cells responsible for IL-4 and IL-5 production are present as precursors that require priming before they develop into lymphokine-secreting cells. J Immunol 1988;141:3445–3455.
8 Swain SE, Weinberg AD, English M: CD4$^+$ T cell subsets: lymphokine secretion of memory cells and of effector cells which develop from precursors in vitro. J Immunol 1989;144:1788–1799.
9 Gajewski TF, Fitch FW: Anti-proliferative effect of IFN-γ in immune regulation. I. IFN-γ inhibits the proliferation of TH2 but not TH1 murine HTL. J Immunol 1988;140:4245–4252.
10 Fernandez-Botran R, Sanders VM, Mosmann TR, Vitetta ES: Lymphokine mediated regulation of the proliferative response of clones of T helper 1 and T helper 2 cells. J Exp Med 1988;168:543–558.

11 Greenbaum LA, Horowitz JB, Woods A, Pasqualina T, Reich EP, Bottomly K: Autocrine growth of CD4⁺ T cells. Differential effects of IL-1 helper and inflammatory T cells. J Immunol 1988;140:1555–1560.

12 Florentino DF, Zlotnik A, Vieira P, Mosmann TR, Howard M, Moore KW, O'Garra A: IL-10 acts on the antigen-presenting cell to inhibit cytokine production by TH1 cells. J Immunol 1991;146:3444–3451.

13 Gajewski TF, Joyce J, Fitch FW: Anti-proliferative effect of IFN-γ in immune regulation. III. Differential selection of TH1 and TH2 murine helper T lymphocyte clones using recombinant IL-2 and recombinant IFN-γ. J Immunol 1988;143:15–22.

14 Magilary BB, Fitch FW, Gajewski TF: Murine hepatic accessory cells support the proliferation of TH1 but not TH2 helper T lymphocyte clones. J Exp Med 1989;170:985–990.

15 Daynes RA, Araneo BA: Contrasting effects of glucocorticoids on the capacity of T cells to produce the growth factors interleukin-2 and interleukin-4. Eur J Immunol 1989;19:2319–2325.

16 Scott DE, Gause WC, Finkelman FD, Steinberg AD: Anti-CD3 antibody induces rapid expression of cytokine genes in vivo. J Immunol 1990;145:2183–2188.

17 Andersson V, Andersson J, Lindfors A, Wagner K, Moller G, Heusser CH: Simultaneous production of interleukin-2, interleukin-4, and interferon–γ by activated human blood lymphocytes. Eur J Immunol 1990;20:1591–1596.

18 Kelly EAB, Cruz ES, Hauda KM, Wassom DL: IFN-γ and IL-5 producing cells compartmentalize to different lymphoid organs in *Trichinella spiralis*-infected mice. J Immunol 1991;147:306–311.

19 Weinberg AD, English ME, Swain SL: Distinct regulation of lymphokine production is found in fresh versus in vitro primed murine helper cells. J Immunol 1990;144:1800–1807.

20 Araneo BA, Dowell TA, Moon HB, Daynes RA: Regulation of murine lymphokine production in vivo: ultraviolet radiation exposure depresses IL-2 and enhances IL-4 production by T cells through an IL-1-dependent mechanism. J Immunol 1989;143:1737–1744.

21 Gahring L, Baltz M, Pepys MB, Daynes R: Effect of ultraviolet radiation on production of epidermal cell thymocyte-activating factor/interleukin-1 in vivo and in vitro. Proc Natl Acad Sci USA 1984;81:1198–1202.

22 Besedovsky H, Del Rey A, Sorkin E, Dinarello CA: Immunoregulatory feedback between interleukin-1 and glucocorticoid hormones. Science 1986;233:652–654.

23 Regelson W, Loria R, Kalimi M: Hormonal intervention; in Pierpaoh W, Spector N (eds): 'Buffer Hormones' or 'State Dependency'. Ann NY Acad Sci 1988;521:260–273.

24 Gordon GB, Shantz LM, Talahy P: Modulation of growth differentiation and carcinogenesis by dehydroepiandrosterone. Adv Enzyme Regul 1986;26:355–382.

25 Daynes RA, Dudley DJ, Araneo BA: Regulation of murine lymphokine production in vivo. II. Dehydroepiandrosterone is a natural enhancer of interleukin-2 synthesis by helper T cells. Eur J Immunol 1990;20:793–802.

26 Meikle AW, Dorchuck RW, Araneo BA, Stringham JD, Evans TG, Spruance SL, Daynes RA: Dehydroepiandrosterone specific receptor binding complex and interleukin-2 stimulation in murine T cells. J Steroid Biochem Molec Biol 1991, in press.

27 Daynes RA, Araneo BA, Dowell TA, Huang K, Dudley D: Regulation of murine lymphokine production in vivo. III. The lymphoid tissue microenvironment exerts regulatory influences over helper T-cell function. J Exp Med 1990;171:979–996.

28 Araneo BA, Dowell T, Terui T, Diegel M, Daynes RA: Dihydrotestosterone exerts a depressive influence on the production of IL-4, IL-5, and γIFN, but not IL-2 by activated murine cells. Blood 1991;78:688–699.

29 Araneo BA, Dowell TA, Terui T, Daynes RA: 1,25-Dihydroxyvitamin D_3 augments the production of IL-4 and IL-5 when T cells are activated under serum-free conditions (submitted).

30 Daynes RA, Meikle AW, Araneo BA: Locally active steroid hormones may facilitate compartmentalization of immunity by regulating the types of lymphokines produced by helper T cells. Res Immunol 1991;142:40–44.

31 Adams JS, Gacad MA: Characterization of 1α-hydroxylation of vitamin D_3 sterols by cultured alveolar macrophages from patients with sarcoidosis. J Exp Med 1985; 161:755–765.

32 Bell NH, Stern PH, Pantzer E, Sinha TK, DeLuca HF: Evidence that increased circulating 1α,25-dihydroxyvitamin D_3 is the probable cause for abnormal calcium metabolism in sarcoidosis. J Clin Invest 1979;64:218–225.

33 Sternberg EM, Hill JM, Chrousos GP, Kamilaris T, Ustwak SJ, Gold PW, Wilder RL: Inflammatory mediator-induced hypothalamic-pituitary-adrenal axis activation is defective in streptococcal cell wall arthritis susceptible Lewis rats. Proc Natl Acad Sci USA 1989;86:2374–2378.

34 Buckler AJ, Vie H, Sonenshein GE, Miller RA: Defective T lymphocytes in old mice. J Immunol 1988;140:2442–2446.

35 Thoman ML, Weigle WO: The cellular and subcellular bases of immunosenescence. Adv Immunol 1989;46:221–261.

36 Araneo BA, Terui T, Dowell T, Daynes RA: Regulation of murine lymphokine production in vivo. IV. Age-associated changes in cytokine production are reversed by DHEAS supplementation (submitted).

37 Betz M, Fox BS: Prostaglandin E_2 inhibits production of Th1 lymphokines but not of Th2 lymphokines. J Immunol 1991;146:108–113.

38 Kelley KW, Brief S, Westly HJ, Novakofski J, Bechtel PF, Simon J, Walter EB: GH_3 pituitary adenoma cells can reverse thymic aging in rats. Proc Natl Acad Sci USA 1986;83:5663–5667.

39 Hartmann DP, Holaday JW, Bernton EW: Inhibition of lymphocyte proliferation by antibodies to prolactin. FASEB J 1989;3:2194–2202.

40 LaMarre J, Wollenberg GK, Gonias SL, Hayes MA: Cytokine binding and clearance properties of proteinase-activated α-macroglobulins. Lab Invest 1991;65:3–14.

41 Fernandez-Botran R: Soluble cytokine receptors: Their role in immunoregulation. FASEB J 1991;5:2567–2574.

42 Daynes RA, Dowell T, Araneo BA: Platelet-derived growth factor is a potent biologic response modifier of T cells. J Exp Med 1991;174:1323–1333.

43 Hart CE, Bailey M, Curtis DA, Osborn S, Raines E, Ross R, Forstrom JW: Purification of PDGF-AB and PDGF-BB from human platelet extracts and identification of all three PDGF dimers in human platelets. Biochemistry 1990;29:166–172.

44   Shimokado K, Raines EW, Madtes DK, Barrett TB, Benditt EP, Ross R: A
     significant part of macrophage-derived growth factor consists of at least two forms of
     PDGF. Cell 1985;43:277–286.
45   Hart CD, Bowen-Pope DF: Platelet-derived growth factor receptor: Current views of
     the two-subunit model. J Invest Dermatol 1990;94:53s–57s.
46   Ferns GAA, Sprugel KH, Seifert RA, Bowen-Pope DF, Kelly JD, Murray M, Raines
     EW, Ross R: Relative platelet-derived growth factor receptor subunit expression
     determines cell migration to different dimeric forms of PDGF. Growth Factors
     1990;3:315–324.

Dr. Raymond A. Daynes, 5C124 Pathology, University of Utah School of Medicine,
50 North Medical Drive, Salt Lake City, Utah 84132 (USA)

Coffman RL (ed): Regulation and Functional Significance of T-Cell Subsets.
Chem Immunol. Basel, Karger, 1992, vol 54, pp 21–43

T Helper Cell Immune Dysfunction in Asymptomatic, HIV-1-Seropositive Individuals: The Role of TH1-TH2 Cross-Regulation

Gene M. Shearer, Mario Clerici [1]

Experimental Immunology Branch, National Cancer Institute,
National Institutes of Health, Bethesda, Md., USA

Introduction

The acquired immunodeficiency syndrome (AIDS) was discovered in 1981 in individuals who were unexpectedly being diagnosed with opportunistic infections [1–3]. This syndrome was subsequently found to result from infection by a human retrovirus, which was eventually termed human immunodeficiency virus type 1 (HIV-1) [4–9]. Unique features of HIV-1 infection include binding to the CD4 molecule [10, 11] and the ultimate depletion of CD4$^+$ T lymphocytes [1–3, 12], with the result that T helper cell (TH) function is lost. Thus, peripheral blood leukocytes (PBL) from AIDS patients are severely depleted of CD4$^+$ T cells, and are unresponsive to a number of in vitro stimuli that normally activate TH [12]. Furthermore, AIDS patients are frequently negative by delayed-type hypersensitivity (DTH) reactions [1–3, 12]. Thus, it is generally considered that HIV-1 infection results in CD4$^+$ TH depletion, which in turn is responsible for susceptibility to opportunistic agents, absence of DTH in vivo and loss of in vitro TH function assessed by T cell proliferation and/or interleukin-2 (IL-2) production.

However, TH defects can be demonstrated even in the absence of a numerical deficiency in CD4$^+$ T lymphocytes. Thus, PBL from AIDS patients, even when enriched for CD4$^+$ cells, failed to respond to tetanus toxoid

[1] The authors express their gratitude to Ms. Janice C. Davis and Ms. Lawann A. Dabney for preparing the manuscript and to Drs. Alan Sher and Herbert C. Morse III, NIAID, NIH, for reviewing the manuscript.

(TET) [13]. Furthermore, a number of laboratories have demonstrated that in vitro TH function can be drastically reduced or lost years before AIDS symptoms start to develop, and before CD4 cell numbers are reduced to critical levels [14–20]. Thus, Smolen et al. [14] and Garbrecht et al. [15] demonstrated defects in the autologous mixed lymphocyte reaction. TH defects in the response to anti-CD3 monoclonal antibody and recall antigens have also been demonstrated in the PBL of asymptomatic, HIV-seropositive (HIV+) individuals by other laboratories [16–20]. This early defect in TH function is even more perplexing when one considers the fact that in situ hybridization assays [21] and polymerase chain reaction analysis [22–24] indicate that the frequency of HIV-infected PBL is not greater than 1/1,000. Thus, the early defect in TH function is not only not correlated with CD4+ cell number, but is also not correlated with the proportion of HIV-infected lymphocytes.

Early T Helper Cell Dysfunction

It should be noted that the early T cell defect of asymptomatic individuals (detected by T cell proliferation and/or IL-2 production) is detected in TH responses to autologous and recall (REC) antigens and anti-CD3, but not in responses to HLA alloantigens (ALLO), or to the T cell mitogen phytohemagglutinin (PHA) [14–20]. Thus, HIV+ individuals have been identified whose PBL fail to generate TH responses to recall antigens such as FLU and TET, but respond to ALLO and PHA [16]. In fact, there is a progressive pattern of TH dysfunction such that among more than 400 asymptomatic seropositive individuals whose PBL were tested: 34% responded to REC, ALLO, and PHA (designated + + +); 44% failed to respond to REC but responded to ALLO and PHA (designated – + +); 9% were unresponsive to REC and ALLO, but were responsive to PHA (designated – – +), and 13% were unresponsive to all three stimuli (designated – – –) [18, and Clerici and Shearer, unpubl. observations]. These findings are summarized in table 1. Although HIV+ adults whose PBL fit into the above in vitro functional categories do not predict who will be more susceptible to opportunistic infections, these categories are predictive for a number of AIDS-related events that indicate a progression toward AIDS. First, the pattern is progressive such that individuals who are in a particular category either remain in that category or move to a more unresponsive category during a 1- to 2-year follow-up period. Fewer than 8% of individuals exhibited spontaneous

Table 1. Patterns of in vitro TH responses to different stimuli in asymptomatic, HIV⁺ individuals (> 400 CD4⁺ cells/μl)

Recall antigens	HLA allo-antigens	Mitogen PHA	Fraction positive	Percent positive	Donor category
+	+	+	140/412	34	(+ + +)
−	+	+	181/412	44	(− + +)
−	−	+	39/412	9	(− − +)
−	−	−	52/412	13	(− − −)

Table 2. Patterns of in vitro TH responses to different stimuli in AIDS patients

Recall antigens	HLA allo-antigens	Mitogen PHA	Fraction positive	Percent positive	Donor category
+	+	+	3/74	4	(+ + +)
−	+	+	16/74	22	(− + +)
−	−	+	17/74	23	(− − +)
−	−	−	38/74	51	(− − −)

improved TH function. The changes in TH function are illustrated for 118 HIV⁺ individuals in figure 1. Only 5/118 (4.2%) (donors 49, 90, 93, 106, and 115) exhibited a TH response to more stimuli at the second test than at the first test. In contrast, 41/118 (35%) exhibited a loss in TH response to one or more stimuli during this period. Furthermore, table 2 illustrates that a higher proportion of AIDS patients fall into the − − + and − − − categories than do asymptomatic HIV⁺ individuals (compare with table 1). Second, the TH functional category of the HIV⁺ individual is predictive for decline in CD4⁺ cell numbers during the subsequent 12–15 months [25]. A + + + functional category was associated with a stable CD4 cell number, whereas the − + +, − − + and − − − categories were all predictive for a significant decline in CD4 cell numbers [25]. Third, among pediatric HIV⁺ patients with AIDS symptoms, we observed a correlation between a particular functional category and susceptibility to opportunistic or bacterial infections [26]. These observations are all consistent with the progressive loss of TH function in PBL of HIV⁺ individuals being an in vitro predictor of immune dysregulation and progression to AIDS.

	Time		Functional Category (IL-2 Production)
0		**Week 62 (± 14 Wks)**	
1 2 3 4 5 6 7 8 9 10 11 12 13 14 15 16 17 18 19 20 21 22 23 24 25 26 27 28 29 30 31 32 33 34 35 36 37 38 39 40 41 42 43 44 45 46		2 3 4 7 11 13 14 16 17 18 20 21 22 25 27 28 31 32 35 38 39 43 44 46 <u>49</u> <u>90</u> <u>93</u>	**+/+/+**
47 48 49 50 51 52 53 54 55 56 57 58 59 60 61 62 63 64 65 66 67 68 69 70 71 72 73 74 75 76 77 78 79 80 81 82 83 84 85 86 87 88 89 90 91 92 93 94 95 96 97		<u>1</u> <u>5</u> <u>6</u> <u>8</u> <u>9</u> <u>12</u> <u>19</u> <u>24</u> <u>26</u> <u>29</u> <u>34</u> <u>36</u> <u>37</u> <u>40</u> <u>41</u> <u>42</u> <u>45</u> 47 48 52 53 55 56 58 59 60 61 62 63 65 66 68 69 70 71 73 74 79 81 82 86 87 88 89 91 94 97 <u>106</u>	**-/+/+**
98 99 100 101 102 103 104 105 106		<u>10</u> <u>15</u> <u>23</u> <u>30</u> <u>50</u> <u>51</u> <u>54</u> <u>57</u> <u>64</u> <u>75</u> <u>77</u> <u>83</u> <u>84</u> <u>85</u> <u>96</u> 98 99 100 101 102 104 105 <u>115</u>	**-/-/+**
107 108 109 110 111 112 113 114 115 116 117 118		<u>33</u> <u>67</u> <u>72</u> <u>76</u> <u>78</u> <u>80</u> <u>92</u> <u>95</u> <u>103</u> 107 108 109 110 111 112 113 114 116 117 118	**-/-/-**

Fig. 1. Changes in TH cell functional status (assessed by IL-2 production by PBL stimulated with recall antigens (REC), HLA alloantigens (ALLO) or phytohemagglutinin (PHA). + and – indicate whether individuals were responsive or unresponsive to the three stimuli: + + + indicates responsive to REC, ALLO, and PHA; – + + indicates unresponsive to REC but responsive to ALLO and PHA; – – – indicates unresponsive to all three stimuli.

Mechanisms Responsible for the Early Loss of TH Function

The finding that TH responses to REC are lost before the responses to ALLO and PHA raises the possibility that the early defect resides in the accessory or antigen-presenting cells (APC). This possibility comes from the fact that TH responses to REC would require both helper cells and APC provided by the HIV$^+$ donor. In contrast, for an ALLO response, APC can be provided by the irradiated allogeneic stimulator PBL from the HIV$^-$ donors used in the study. Furthermore, the accessory cell requirements for PHA are probably less stringent than those for REC and ALLO. Thus, it could be argued that the relative high proportion (44%) of the $-++$ functional category could be the result of an HIV-induced defect in APC function. In fact, it has been well documented that CD4$^+$ cells of the monocyte/macrophage lineage can be infected with HIV-1 in vitro, and that these cells are infected in HIV$^+$ patients [27–35]. Furthermore, defective monocyte/macrophage function has been reported in cells isolated from AIDS patients [36–38]. Such reports have led to the suggestion that an APC defect could exist in HIV$^+$ and/or AIDS patients. Despite the above findings, in vitro experiments to independently test TH and APC function of PBL from HIV$^+$ patients with some symptoms failed to show an APC defect, but did demonstrate a defect in TH function [12, and Clerici et al., submitted for publication]. In these experiments, PBL from monozygotic twins, one of whom was HIV$^+$ and the other of whom was HIV$^-$, were used as the sources of TH and APC for generating in vitro TH responses to REC. It was found that the TH from the HIV$^-$ twin generated TH function when APC function was provided by PBL from the HIV$^+$ twin. However, no TH function was detected when TH from the HIV$^+$ twin were cocultured with APC from the HIV$^-$ twin. The failure to detect a defect in APC function was not due to a flaw in the experimental design, because we have been able to detect defects in APC in approximately 50% of AIDS patients [39], but not in the PBL of asymptomatic HIV$^+$ individuals [40].

The experiments noted above strongly suggest that an early defect resides in the TH themselves. If this interpretation of the data is correct, what is the nature of the defect in TH function? Because we have detected multiple levels of TH dysfunction, the mechanism(s) responsible for the defect may be complex. Thus, a mechanism(s) that accounts for the selective loss of TH function to REC should also consider the progressive loss of responses to ALLO and PHA during the asymptomatic period.

A mechanism that has been suggested to account for the early loss of TH function to REC is the selective depletion of CD4$^+$ cells that express markers characteristic of memory T cells (e.g., CD29$^+$) [41–44]. Thus, it has been demonstrated the reduction in TH function is associated with a selective depletion of CD29$^+$ cells [44]. Furthermore, it has been reported that CD4$^+$ T cells expressing memory markers are preferentially infected with HIV-1 in vitro and that memory T cells constitute a majority of HIV⁺ cells in the PBL of infected patients [43]. These findings are consistent with the observations that activated T cells are more easily productively infected by HIV-1 than resting T cells [45–47].

The interpretation is, however, complicated by the selective loss of TH to REC but not to ALLO in approximately 40% of asymptomatic HIV⁺ individuals [18]. Analysis of the TH function to ALLO indicated that three distinct TH-APC pathways can be demonstrated in the mixed lymphocyte reaction [48]; one involving CD4⁺ TH that recognize HLA alloantigens that have been processed and presented by autologous APC (self-restricted); a second pathway involving CD4⁺ TH that recognize HLA alloantigens (probably class II) directly on allogeneic APC (allo-restricted), and a third pathway that involves CD8⁺ TH that recognize HLA alloantigens (probably class I) directly on allogeneic APC (allo-restricted). The PBL from the HIV⁺ individuals who fail to respond to REC but respond to ALLO, also fail to respond to the self-restricted component of ALLO, but respond to ALLO through the two allo-restricted pathways [49]. Thus, this early defect is seen in self-restricted TH function, which would also include memory responses, and one can question whether the defect is actually for memory or self-restricted responses. Also, the fact that the allo-restricted CD4-mediated component of the TH response to ALLO is intact indicates that this early defect does not involve CD4⁺ TH per se.

A number of mechanistic questions remain unresolved. For example, is the same mechanism that is responsible for the loss of TH function to REC also responsible for the subsequent loss of TH function to ALLO and PHA? Could the greater susceptibility to REC responses be due to differences in TH precursor frequencies with the precursors to REC being less frequent, and therefore the REC responses more drastically affected than the responses to ALLO or PHA?

Could the early loss of TH function be due to some form of active immune suppression? Although suppression was considered as an important mechanism of immune regulation in the 1970s [50], this phenomenon has become less popular during the 1980s. Nevertheless, several laboratories have tested whether active suppression or down-regulation of immune responsiveness could contribute to the loss of T-cell immunity in HIV⁺ patients. Thus, soon after the AIDS epidemic was discovered, Laurence et al. [51, 52] demonstrated that PBL from AIDS patients could suppress T-cell responses generated by PBL from HIV⁻ donors, and that the patients' PBL could produce a factor that suppressed the responses of PBL from the HIV⁻

donors. Other studies demonstrated that PBL from AIDS patients could suppress the T-cell proliferative [53, 54] or cytotoxic T-lymphocyte [55] responses of PBL from uninfected donors. In order to avoid potential problems introduced by the effects of coculturing allogeneic PBL [56], we have used PBL from monozygotic twins, one of whom was HIV⁺ and the other of whom was HIV⁻, as well as cryopreserved PBL from the longitudinally studied individual, taken before HIV infection and 7 years later, after AIDS developed [Clerici et al., submitted for publication]. Our results indicate that CD8⁺ T cells (and possibly monocytes) from the HIV⁺ twin suppressed self-restricted, CD4-mediated TH function, but not allo-restricted TH function nor PHA responses. Furthermore, a soluble factor was produced that also suppressed only self-restricted TH responses. This factor was found not to be infectious HIV [Clerici et al., submitted for publication].

Any of the several different soluble factors could be responsible for the suppressive effect, including HIV products and cytokines. For example, the gp120 of the envelope of HIV-1 has been found to be immunosuppressive when added to cultures of PBL from HIV⁻ individuals [57–60]. Furthermore, gp120 and anti-gp120 antibody can synergize to inhibit T-cell activation [61]. Autoantibodies generated against the gp41 of HIV-1 cross-react with a nonpolymorphic determinant of HLA class II, and can block T-cell responses in vitro [62, 63]. The tat protein is an HIV product that has been demonstrated to selectively inhibit T-cell proliferative responses to recall antigens [64]. Transforming growth factor β (TGF-β) could also contribute to the loss of TH function because the elevated levels of TGF-β seen in HIV⁺ patients correlate with impaired proliferative responses to recall antigens, and the lack of response can be reversed by anti-TGF-β antibody [65]. The reports summarized above indicate that a number of soluble products resulting from HIV-1 infection could contribute to the TH defect(s) seen in asymptomatic, HIV⁺ individuals.

T Helper Cell Functional Analysis in Clinical Trials

A number of phase I protocols have been performed and phase II protocols are in progress to test the efficacy of antiretroviral drug therapy such as AZT [66], 2′,3′-dideoxyinosine (ddI) [67, 68], and 2′3′-dideoxycytidine (ddC) [69] in reducing AIDS mortality and retarding the progression to AIDS. Several surrogate markers including the CD4:CD8 ratio, CD4 cell number, p24 antigen level, HIV-1 viral load serum neoptrin level, and β_2-microglobulin levels have been used to assess the efficacy of antiretroviral therapy. However, none of these markers have been completely satisfactory, and most do not assess immune function. The CD4 cell count has been considered to be the most relevant of the surrogate markers. However, as noted above, the CD4 cell number is not necessarily correlated with TH function, and changes in CD4 cell number are often minimal and may require several months of therapy to be detected. Therefore, a surrogate marker is needed that is sensitive enough to detect antiretroviral-induced changes earlier. Because there is a progressive loss in TH function that involves responses to REC, ALLO, and PHA (see above), we have used this in vitro test and the possible reversal of loss of TH function to assess antiretroviral therapy. We observed that PBL from more than 75% of adult patients on AZT and more than 50% or children on ddI exhibited improved TH function during therapy [Clerici et al., submitted for publication]. Some of the patients exhibited improved TH function within 8 weeks of initiating therapy, and improved TH function continued to be evident more than 2 years into therapy. These relatively rapid antiretroviral-induced immuno-logic improvements in TH function also suggest that soluble factors contribute to immune dysfunction in these patients, as does the report that AIDS-associated cognitive function can be partially reversed by ddI therapy [70].

Possible Role of TH1-TH2 Cross-Regulation in the
HIV-Induced Early TH Defects

Some of the findings summarized above suggest that soluble factors contribute to the loss of TH function in asymptomatic, HIV+ individuals, assessed by proliferation and IL-2 production. Possible candidates for loss of these functions include one or more cytokines. Cytokines such as TNF-α and IL-6 have been demonstrated to be important in regulating HIV-1 infection [71]. Cytokines could also contribute to the T-cell immune dysregulation that results from HIV-1 infection.

In the mid- to late 1980s, Mosmann et al. [72–74] isolated two distinct groups of murine TH clones: one type of clone, designated TH1, produced IFN-γ and IL-2 but not IL-4 or IL-5; the second type of clone, designated TH2, produced IL-4 and IL-5 but not IFN-γ or IL-2. Other clones were identified that produced all of the above mentioned cytokines, as well as IL-3 and GM-CSF, and were designated THO [75, 76]. CD4+ T cells from unimmunized mice produced only IL-2, and clones derived from these cells have been designated THP [75–77]. In general, TH1 clones mediate helper cell functions associated with cytotoxic T lymphocytes and DTH reactions, whereas TH2 clones are more effective at stimulating B cells and antibody production, particularly of the IgM, IgG1, IgA and IgE classes [78–80]. These rules are not absolute, however, because TH1 clones can stimulate the production of complement-fixing IgG2a antibodies under certain conditions [79]. It appears that the murine immune system frequently chooses between cell-mediated responses (TH1) and antibody production (TH2), and that these two types of responses can be reciprocally controlled [81, 82]. Thus the induction of a TH1-mediated response can lead to suppression of TH2 function, and vice versa. Cross-regulation of TH1 and TH2 types of responses has been recently demonstrated for mice. Thus, IFN-γ produced by TH1 clones inhibits the cytokines produced by TH2 clones, and IL-10 (formerly known as cytokine synthesis inhibitory factor or CSIF) produced by TH2 clones down-regulates the cytokines produced by TH1 clones [81, 82].

Until recently, the TH1 and TH2 profiles reported for mice did not appear to extend to man. However, it has now been demonstrated by Romagnani and co-workers [83, 84] that T lymphocytes taken from patients who have autoimmune diseases, or infections, or from individuals who have been immunized, can be cloned, and that the clones produce cytokines that are characteristic of the murine TH1 and TH2 types. Papers continue to appear in the literature which support the concept that human helper cells can also be subdivided into TH1 and TH2 functional categories [85, 86]. In fact, it has been reported that cloned CD4 cells from AIDS patients generated reduced amounts of IL-2 and IFN-γ but elevated levels of B cell growth and differentiation factors compared with CD4 clones from healthy control donors [87].

We recently began to consider the possibility that TH1-TH2 cross-regulation may play an important role in HIV-1 infection and/or progression to AIDS. This possibility is based on the following observations. First, HIV-1 infection results in B-cell activation and hypergammaglobulinemia [88–90], including increases in levels of IgA [91] and IgE [92]. Levels of IL-6 can also be elevated in HIV+ patients [93, 94], and non-Hodgkin's B-cell lymphomas are not uncommon in AIDS [95]. These observations indicate that despite the immunosuppressive characteristic of HIV infection and AIDS development, immune regulatory pathways that control B-cell function appear to be overstimulated. Second, the PBL of HIV-infected individuals can lose the ability to produce IL-2 in response to antigenic stimulation several years before symptoms develop [18]. This observation indicates that TH activity for a TH1 function is lost, despite enhanced immunologic activity for a TH2-controlled function (B-cell activation).

Third, hundreds (possibly thousands) of individuals have been exposed to HIV-1, but have not seroconverted. Recently, we began to study at-risk individuals who fit into this category by testing the ability of their PBL to generate in vitro TH responses (by proliferation and IL-2 production) to four synthetic peptides which were found to be immunogenic in mice immunized with gp160 [96–99]. The peptides stimulated TH responses in PBL of seropositive humans [100] and human volunteers immunized with a recombinant gp160 candidate AIDS vaccine [101], but did not stimulate TH responses of PBL from more than 120 unimmunized or low-risk seronegative individuals [100, 101, Clerici et al., unpubl. observations]. The at-risk groups that we have studied thus far include homosexual men and intravenous drug users who have continued high-risk practices, as well as health care workers who have had accidental exposures to HIV-1 but have not seroconverted. We observed that the PBL from a high proportion of seronegative homosexual men (9/12) [102, 103] and 50% of drug abusers exhibited HIV-specific TH activity when stimulated with the HIV-1 synthetic peptides, and that PBL from several seronegative exposed health care workers generated in vitro TH responses to these peptides [Clerici and Shearer, unpubl. observations]. PBL from the first six of the seronegative, T-cell-responsive homosexual men studied were found to be negative for HIV infection by the polymerase chain reaction (PCR) [102, 103]. Two of these individuals subsequently and simultaneously seroconverted and PCR-converted, one after 19 months and the other after 6 months into the study. Our observations of HIV-specific TH+ reactivity in the absence of HIV antibodies raises the possibility that cell-mediated immunity (TH1) may be associated with: (a) protection against HIV-1 infection in exposed individuals, and/or (b) regulation of the magnitude of the viral infection in individuals who have been HIV-1-infected. In contrast, seroconversion could indicate that sufficient viral replication has occurred to activate the B-cell component (TH2) of immunity. In this context, it should be noted that in animal studies, a low dose of antigen was shown to activate cell-mediated immunity (possibly TH1-like), whereas a higher dose of antigen elicited antibody responses [104–106]. Furthermore, low-dose immunization of HIV-seronegative human volunteers with the recombinant gp160 candidate AIDS vaccine (40–80 µg/recipient) resulted in only HIV-specific cell-mediated immunity (proliferation, IL-2 production, and cytotoxic T lymphocytes), but not in antibody production [101], whereas higher doses of immunization resulted in both cellular and humoral immunity [107]. Thus, it is possible that in individuals who have been naturally exposed to HIV-1, HIV-specific TH immunity in the absence of antibody (TH1) is indicative of low viral load, whereas HIV seroconversion is indicative of viral replication and activation of TH2 cells. In other words, it is possible that TH1 immunity may be associated with immune protection, whereas TH2 immunity may be indicative of viral replication that is not any longer controlled by the immune system.

Fourth, AIDS has reached epidemic proportions in a number of tropical areas of the world such as in Asia, Latin America and Africa [108, 109]. It is in these same regions that helminth parasite infections are prevalent. In murine models it has been clearly demonstrated that helminth parasitic infections can have dramatic effects on TH1-TH2 cross-regulation [82, 110–112]. For example, *Nippostrongylus* infection results in increases of IL-4, IL-5 and IL-6, but reduced levels of IFN-γ and IL-2 [76]. Furthermore, BALB/c mice are susceptible to *Leishmania major*, and produce high levels of IL-4 and IgE, but low levels of IFN-γ and a weak DTH. In contrast, C57BL/6 mice are resistant to *L. major*, produce high levels of IFN-γ and a strong DTH, but low levels of IL-4 and IgE. Infection of mice with *Schistosoma mansoni* that resulted in an egg-producing phase was associated with disease and a TH2 pattern of cytokine production such as IL-4 and IL-5, whereas vaccination with attenuated larva was protective and associated with a TH1 pattern of cytokine production such as IFN-γ and IL-2 [112]. It has also been demonstrated in the murine model of *S. mansoni* that infection can begin predominantly as a resistant TH1 response, and as infection progresses, TH function can convert to a TH2 type of response [110, 113]. These and other studies suggest that for some infections, TH1 function provides immune protection, whereas TH2 function is not protective. Is it possible that TH1 immunity is protective against HIV-1 infection and/or progression of infected individuals to AIDS? Could helminth parasitic infection in a human population result in a predominance of TH2 over TH1, thereby making the population more susceptible to HIV-1 infection and/or more rapid progression to AIDS? Would such helminth infections or any immune stimulus that would induce a TH1-to-TH2 switch serve as a cofactor in AIDS susceptibility?

If TH1-TH2 cross-regulation contributes to the resistance and susceptibility to HIV infection and/or AIDS progression, it should be of interest to compare TH1 and TH2 cytokine production by PBL from asymptomatic,

Table 3. Comparisons of IL-2 and IL-4 production by PBL from HIV$^-$ and asymptomatic HIV$^+$ individuals

HIV status	TH functional category	Relative ratio of IL-2:IL-4
–	REC$^+$, ALLO$^+$, PHA$^+$	5.6
+	REC$^+$, ALLO$^+$, PHA$^+$	2.2
+	REC$^-$, ALLO$^+$, PHA$^+$	0.1
+	REC$^-$, ALLO$^-$, PHA$^+$	0.3

HIV$^+$ individuals with the cytokine patterns by PBL from low-risk HIV$^-$ individuals. We have begun such a study and have found that REC-stimulated IL-2 production is high and PHA-stimulated IL-4 production is low in the HIV$^-$ control group. In contrast, HIV$^+$ individuals exhibit a more complex pattern of cytokine production in which: (a) the most functionally asymptomatic (+ + +) HIV$^+$ individuals still showed a predominance of IL-2 over IL-4, but the IL-2 was lower and IL-4 higher than in HIV$^-$ controls; (b) individuals who had lost TH function for REC only (– + +) showed a reversal in IL-2 and IL-4 production that was due to both a reduction in IL-2 *and* to an increase in IL-4 production, and (c) HIV$^+$ individuals who had also lost the ability to respond to ALLO (– – +) exhibited a reduction in both IL-2 and IL-4 production. This pattern of cytokine production is summarized in table 3, and demonstrates that: (a) the TH function that we have assessed in asymptomatic HIV $^+$ individuals [18] not only involves a reduction in IL-2 production, but also a concomitant increase in IL-4 production, and (b) that progressive loss in TH function that we have seen, assessed by the sequential loss of IL-2 production to different stimuli [18], is also reciprocally reflected in IL-4 production.

On the basis of the four points considered above, as well as the summary of our findings in table 3, we propose the following model of HIV-1 resistance and susceptibility and/or AIDS progression which is dependent on TH1-TH2 cross-regulation:

(1) HIV-specific TH1 responses are immunoprotective and prevent infection, and/or keep the infection under control. Limiting amounts of HIV-1 that induce T-cell immunity but not antibody production are characteristic of this phase.

(2) HIV seroconversion represents a state in which the viral replication has increased and has generated enough antigen to activate TH2 responses

and HIV-specific antibodies. Although antibodies provide some immune protection, a state of seropositivity is ultimately indicative of the failure of the immune system to keep the virus in check.

(3) An individual can be in a 'predominant TH1' (TH1 > TH2) or a 'predominant TH2' (TH2 > TH1) state, depending on exposure to recent immunologic stimuli that may be unrelated to HIV-1 infection.

(4) An individual in a TH1 state is more resistant to HIV-1 infection and/or AIDS progression than an individual who is in a TH2 state, and can hold the virus in check, unless or until that individual moves to a TH2 state, possibly due to unrelated immunologic events that promote a TH1-to-TH2 switch.

(5) IL-10 is an important cytokine in the regulation of resistance and susceptibility to HIV-1 infection and/or progression to AIDS, and increases in IL-10 will be predictive for a more susceptible state. Any non-HIV infection or immune stimulus that enhances IL-10 production will increase susceptibility for infection or AIDS progression.

This model is illustrated in figure 2. In figure 2a, predominance of TH1 over TH2 keeps HIV infection under control. Loss of TH1-mediated protective immunity eventually occurs possibly due to cofactors that induce a progression to AIDS. In figure 2b, TH1 over TH2 prevents infection following HIV exposure by a TH1-mediated mechanism. Loss of TH1-protective immunity, possibly due to cofactors that induce a TH1-to-TH2 switch, renders the individual susceptible to a subsequent exposure to HIV. A model similar to ours has been recently developed by Morse and co-workers [114, 115], based on data obtained using the murine retrovirus-induced lymphoproliferative MAIDS model. These investigators demonstrated that the lymphoproliferative disease resulting in susceptible mice with a mixture of murine leukemia viruses (LP-BM5) is associated with a very early TH0 type of response (with production of IFN-γ, IL-2, IL-4, IL-5, IL-6, and IL-10), followed by TH1 predominance (IFN-γ and IL-2) and, later when B-cell hyperplasia was evident, by TH2 predominance (IL-4, IL-5, IL-6 and IL-10). It is also noteworthy that BALB/c mice immunized with sperm whale myoglobin 2 months after infection with *S. mansoni* exhibited a shift in myoglobin-specific TH1 cytokine production (IFN-γ and IL-2) toward a TH2-type response (IL-4) [116].

IL-10, which induces a TH1-to-TH2 switch in mice [81], will be a critical cytokine to assess individuals in high-risk groups, as well as in HIV+ individuals whose infection appears to be latent. It would be of interest to assess some parameter of the relative amounts of TH1 and TH2 cytokine

Models of TH1 <---> TH2 cross-regulation and AIDS

Controlling AIDS progression:

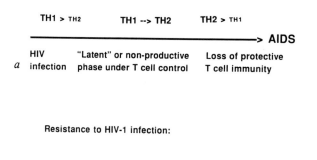

Resistance to HIV-1 infection:

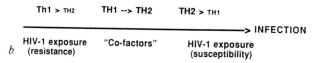

Fig. 2. Two possible models for the role of TH1-TH2 cross-regulation in resistance and susceptibility to HIV-1 infection and/or progression to AIDS. *a* Diagrams on a time line: the regulation of progression to AIDS which would be dependent on the relative strengths of TH1 and TH2 types of immunity. *b* Diagrams on a time line: resistance and susceptibility to HIV-1 infection which would be dependent on the relative strengths of TH1 and TH2 types of immunity.

production in these individuals, as well as IL-10. It is also possible that these cytokine profiles will be abnormal in those regions of the world in which AIDS has reached epidemic proportions. It should be noted, however, that it is not yet clear that human IL-10 controls TH1-TH2 cross-regulation in the same way that murine IL-10 does.

If IL-10-controlled TH1-TH2 cross-regulation is important in resistance and susceptibility to HIV-1 infection and/or AIDS progression, consideration should be given to the possibility of immune-based therapy that would be designed to maintain an individual in a predominant TH1 state, and to convert an infected or at-risk individual who is in a predominant TH2 state to a predominant TH1 state. Furthermore, AIDS vaccine programs should take into consideration TH1-TH2 cross-regulation. This could be important both for the HIV-1 antigenic determinants used to immunize and the adjuvant employed. For example, in mice some antigenic determinants of *L. major* induce TH1 responses, whereas others can induce TH2 responses [111]. Thus, one might want to consider HIV-1 subunit vaccines that would primarily induce cell-mediated rather than humoral immunity. This is different than the criteria now being used to evaluate candidate AIDS vaccines, which are either anti-HIV antibody production only or stimulation of both humoral and cellular immunity [107]. Furthermore, certain adjuvants are primarily TH1 stimulators, whereas others can be TH2 activators. Alum, the adjuvant currently used in AIDS vaccine trials, predominantly activates TH2 responses in mice, whereas BCG is a potent TH1 adjuvant [T.R. Mosmann and B.R. Bloom, pers. commun.].

If our model is correct, AIDS researchers may have been studying the immunologic failures of HIV-1 infection for the past decade (seropositive individuals), and ignoring what may be the immunologic successes of HIV-1 exposure (exposed seronegative individuals who exhibit evidence of TH immunity to HIV antigens). Investigators in AIDS research need to become more familiar not only with the cytokines that regulate HIV-1 infection, but also with those that regulate T helper cell immunity. Based on the findings summarized in this article, we consider that IL-10 and possibly other cytokines that regulate TH1-TH2 cross-regulation will be important factors for controlling the AIDS epidemic, particularly during the early phases of exposure and infection.

References

1 Gottlieb MS, Shraff R, Schanker HM, Weisman JD, Fan PT, Wolf RA, Saxon A: *Pneumocystis carinii* pneumonia and mucosal candidiasis in previously healthy homosexual men. N Engl J Med 1981;305:1425–1430.
2 Masur H, Michelis MA, Greene JB, Onoratol I, Vande Stouwe RA, Holzman RS, Wormser G, Brettman L, Lange M, Murray HW, Cunningham-Rundles S: An outbreak of community-acquired *Pneumocystis carinii* pneumonia. N Engl J Med 1981;305:1431–1438.
3 Siegal FP, Lopez C, Hammer GS, Brown AE, Kornfeld SJ, Gold J, Hassett J, Hirschman SZ, Cunningham-Rundles C, Adelsberg J, Parham DM, Siegel M, Cunningham-Rundles S, Armstrong D: Severe acquired immunodeficiency in male homosexuals manifested by chronic penianal ulcerative herpes simplex lesions. N Engl J Med 1981;305:1439–1444.
4 Barre-Sinoussi F, Chermann JC, Rey F, Nugeyre MT, Chamaret S, Gruest J, Dauguet C, Axler-Blin C, Vesinet-Brun F, Rouzioux C, Rozenbaum W, Montagnier

L: Isolation of a T lymphotropic retrovirus from a patient at risk for acquired immune deficiency syndrome (AIDS). Science 1983;220:868–871.

5 Popovic M, Sarngadharan MG, Read E, Gallo RG: Detection, isolation, and continuous production of cytopathic retroviruses (HTLV-III) from patients with AIDS and pre-AIDS. Science 1984;224:497–500.

6 Gallo RC, Salahuddin SZ, Popovic M, Shearer GM, Kaplan M, Haynes BF, Palker TJ, Redfield R, Oleske J, Safai B, White G, Foster P, Markham PD: Frequent detection and isolation of cytopathic retroviruses (HTLV-III) from patients with AIDS and at risk for AIDS. Science 1984;224:500–503.

7 Schupbach J, Popovic M, Gilden RV, Gonda MA, Sarngadharan MG, Gallo RC: Serological analysis of a subgroup of human T-lymphotropic retroviruses (HTLV-III) associated with AIDS. Science 1984;224:503–505.

8 Sargadharan MG, Popovic M, Bruch L, Schupback J, Gallo RC: Antibodies reactive with a human T-lymphotropic retrovirus (HTLV-III) in the serum of patients with AIDS. Science 1984;224:506–508.

9 Levy JA, Hoffman AD, Kramer SM, Landis JA, Shimabukuro JM: Isolation of lymphocytopathic retroviruses from San Francisco patients with AIDS. Science 1984;225:840–842.

10 Dalgleish AG, Beverly PCL, Clapham PR, Crawford DH, Greaves MF, Weiss RA: The CD4 (T4) antigen is an essential component of the receptor for the AIDS retrovirus. Nature 1984;312:763–767.

11 Klatzmann D, Champagne E, Chamaret S, Gruest J, Guetard D, Hercend T, Gluckman JC, Montagnier L: T-lymphocyte T4 molecule behaves as the receptor for human retrovirus LAV. Nature 1984;312:767–768.

12 Fauci AS: The human immunodeficiency virus: Infectivity and mechanisms of pathogenesis. Science 1988;239:617–623.

13 Lane HC, Depper JM, Greene WC, Whalen G, Waldmann TA, Fauci AS: Qualitative analysis of immune function in patients with the acquired immunodeficiency syndrome. Evidence for a selective defect in soluble antigen recognition. N Engl J Med 1985;313:79–84.

14 Smolen JS, Bettleheim P, Koller U, McDougal S, Graninger W, Luger TA, Knapp W, Lechner K: Deficiency of the autologous mixed lymphocyte reaction in patients with hemophilia treated with communical factor VIII concentrate: correlation with T cell subset distribution, antibodies to lymphadenopathy-associated or human T lympho-tropic virus, and analysis of the basis of the deficiency. J Clin Invest 1985;75:1828–1834.

15 Garbrecht FC, Siskind GW, Wexler ME: Lymphocyte transformation induced by autologous cells. XVIII. Impaired autologous mixed lymphocyte reaction in subjects with AIDS-related complex. Clin Exp Immunol 1987;67:245–251.

16 Shearer GM, Bernstein DC, Tung KS, Via CS, Redfield R, Salahuddin SZ, Gallo RC: A model for the selective loss of major histocompatibility self-restricted T-cell immune responses during the development of acquired immune deficiency syndrome (AIDS). J Immunol 1986;137:2514–2521.

17 Miedema F, Chantal Petit AJ, Terpstra FG, Eeftinek Schattenkerk JKM, deWolf F, Al BJM, Roos M, Lange JMA, Danner SA, Goudsmit J, Schellekens PTA: Immuno-logical abnormalities in human immunodeficiency virus (HIV)-infected asymptom-atic homosexual men: HIV affects the immune system before CD4[+] T cell depletion occurs. J Clin Invest 1988;82:1908–1914.

18 Clerici M, Stocks NI, Zajac RA, Boswell RN, Lucey DR, Via CS, Shearer GM: Detection of three distinct patterns of T helper cell dysfunction in asymptomatic, human immunodeficiency virus-seropositive patients. J Clin Invest 1989;84:1892–1899.

19 Schellenkens PT, Roos MT, DeWolfe F, Lange JM, Miedema F: Low T-cell responsiveness to activation via CD3/TCR is a prognostic marker for acquired immunodeficiency syndrome (AIDS) in human immunodeficiency virus-1 (HIV-1)-infected men. J Clin Immunol 1990;10:121–127.

20 Petersen J, Church J, Gomperts E, Parkman R: Lymphocyte phenotypic does not predict immune function in pediatric patients infected with human immunodeficiency virus type 1. J Pediatr 1989;115:944–948.

21 Harper ME, Marselle LM, Gallo RC, Wong-Staal F: Detection of lymphocytes expressing human T-lymphotropic virus type III in lymph nodes and peripheral blood from infected individuals by in situ hybridization. Proc Natl Acad Sci USA 1986;83:772–776.

22 Ho DD, Mougdil MS, Alam M: Quantitation of human immunodeficiency virus type 1 in the blood of infected persons. N Engl J Med 1989;321:1621–1625.

23 Hufert FT, von Laer D, Schramm C, Tarnok A, Schmitz H: Detection of HIV-1 DNA in different subsets of human peripheral blood mononuclear cells using the polymerase chain reaction. Arch Virol 1989;106:341–345.

24 Schnittman SM, Psallidopoulos MC, Lane HC, Thompson L, Baseler M, Massari F, Fox CH, Salzman NP, Fauci AS: The reservoir for HIV-1 in human peripheral blood is a T cell that maintains expression of CD4. Science 1989;245:305–308.

25 Lucey DR, Melcher GP, Hendrix CW, Zajac RA, Goetz DW, Butzin CA, Clerici M, Warner RD, Abbadessa S, Hall K, Jaso R, Woolford B, Miller S, Stocks NI, Salinas CM, Wolfe WH, Shearer GM, Boswell RN: Human immunodeficiency virus infection in the US Air Force: Seroconversions, clinical staging and assessment of a T-helper cell functional assay to predict change in CD4$^+$ T cell counts. J Infect Dis 1991;164:631–637.

26 Roilides E, Clerici M, DePalma L, Rubin M, Pizzo PA, Shearer GM: Helper T-cell responses in children infected with human immunodeficiency virus type 1. J Pediatr 1991;118:724–730.

27 Gendelman HE, Orenstein JM, Baca LM, Weiser B, Burger H, Kalter DC, Maltzer MS: The macrophage in the persistence and pathogenesis of HIV infection. AIDS 1989;3:475–495.

28 Ho DD, Rota TR, Hirsch MS: Infection of monocyte/macrophages by human T-lymphotropic virus type III. J Clin Invest 1986;77:1712–1715.

29 Nicholson JKA, Cross GD, Gallaway CS, McDougal JS: In vitro infection of human monocytes with human T-lymphotropic virus type III/lymphadenopathy-associated virus (HTLV-III/LAV). J Immunol 1986;37:323–329.

30 Gartner S, Marikovits P, Marovitz DM, Kaplan MH, Gallo RC, Popovic M: The role of mononuclear phagocytes in HTLV-III/LAV infection. Science 1986;233:215–219.

31 Koenig S, Gendelman HE, Orenstein JM, DalCanto MC, Pezeshkpour GH, Yungbluth M, Janotta F, Aksamit A, Martin MA, Fauci AS: Detection of AIDS virus in macrophages in brain tissue from AIDS patients with encephalopathy. Science 1986;233:1089–1093.

32 Eilbott DJ, Peress N, Burger H, LaNeve D, Orenstein J, Gendelman HE, Seidman R, Weiser B: Human immunodeficiency virus type 1 in spinal cords of acquired

immunodeficiency syndrome in patients with myelopathy: expression and replication in macrophages. Proc Natl Acad Sci USA;86:3337–3341.

33 Prince HE, Moody DJ, Shubin BI, Fahey JL: Defective monocyte function in acquired immune deficiency syndrome (AIDS): evidence from a monocyte-dependent T-cell proliferative system. J Clin Immunol 1985;5:21–25.

34 Rich EA, Toossi Z, Fujiwara H, et al: Defective accessory function of monocytes in human immunodeficiency virus-related disease syndromes. J Lab Clin Med 1988; 112:174–181.

35 Chantal-Petit AJ, Tersmetle M, Terpstra FG, deGoede REY, vanLier RAW, Miedema F: Decreased accessory cell function by human monocytic cells after infection with HIV. J Immunol 1988;140:1485–1489.

36 Smith PD, Ohura K, Masur H, Lane HC, Fauci AS, Wahl SM: Monocyte function in the acquired immune deficiency syndrome defective chemotaxis. J Clin Invest 1984; 74:2121–2128.

37 Enk C, Gerstoft J, Moller S, Remvig L: Interleukin-1 activity in the acquired immunodeficiency syndrome. Scand J Immunol 1986;23:491–497.

38 Haas JG, Riethmuller G, Ziegler-Heitbrock HWL: Monocyte phenotype and function in patients with the acquired immunodeficiency syndrome (AIDS) and AIDS-related disorders. Scand J Immunol 1987;26:371–379.

39 Clerici M, Landay A, Kessler HA, Zajac RA, Boswell RN, Muluk SC, Shearer GM: Multiple patterns of alloantigen presenting/stimulatory cell dysfunction in patients with AIDS. J Immunol 1991;146:2207–2213.

40 Clerici M, Stocks NI, Zajac RA, Boswell RN, Shearer GM: Accessory cell function in asymptomatic human immunodeficiency virus-infected patients. Clin Immunol Immunopathol 1990;54:168–173.

41 Giorgi JV, Nishanian PG, Schmid I, Hultin LE, Cheng HL, Detels R: Selective alterations in immunoregulatory lymphocyte subsets in early HIV infection. J Clin Immunol 1987;7:140–150.

42 DeMartini RM, Turner RR, Formenti SC, Boone DC, Bishop PC, Levine AM, Parker JW: Peripheral blood mononuclear cell abnormalities and their relationship to clinical course in homosexual men with HIV infection. Clin Immunol Immunopathol 1988;46:258–271.

43 Schnittman SM, Lane HC, Greenhouse J, Justement JS, Baseler M, Fauci AS: Preferential infection of CD4+ T-memory cells by human immunodeficiency virus type I: evidence for a role in the selective T cell functional defects observed in infected individuals. Proc Natl Acad Sci USA 1990;87:6058–6062.

44 Van Noesel CJM, Gruters RA, Terpstra FG, Miedema F: Functional and phenotypic evidence for a selective loss of memory T cells in asymptomatic human immunodeficiency virus-infected men. J Clin Invest 1990;86:293–299.

45 Zagury D, Bernard J, Leonard R, Cheynier R, Feldman M, Sarin PS, Gallo RC: Long-term cultures of HTLV-III-infected T cells: A model of cytopathology of T-cell depletion in AIDS. Science 1985;231:850–853.

46 Folks T, Kelly J, Benn S, Kinter A, Justerment J, Gold J, Redfield R, Sell KW, Fauci AS: Susceptibility of normal lymphocytes to infection with HTLV-III/LAV. J Immunol 1986;136:4049–4053.

47 Gruters RA, Oho SA, Al BJM, Verhoven AJ, Verweij CL, vanLier RAW, Miedema F: Non-mitogenic T cell activation signals are sufficient for induction of human immunodeficiency virus transcription. Eur J Immunol 1991;21:167–172.

48 Via CS, Tsokos G, Stocks NI, Clerici M, Shearer GM: Human in vitro allogeneic responses: demonstration of three pathways of T helper cell activation. J Immunol 1990;144:2524–2528.

49 Clerici M, Via CS, Lucey DR, Roilides E, Pizzo PA, Shearer GM: Functional dichotomy of CD4+ T helper lymphocytes in asymptomatic human immunodeficiency virus infection. Eur J Immunol 1991;21:665–670.

50 Green DR, Flood PM, Gershen RK: Immunoregulatory T-cell pathways. Annu Rev Immunol 1983;1:439–463.

51 Laurence J, Gottlieb AB, Kunkel HG: Soluble suppressor factors in patients with acquired immune deficiency syndrome and its prodrome. Elaboration in vitro by T lymphocyte adherent cell interactions. J Clin Invest 1983;72:2072–2081.

52 Laurence J, Mayer L: Immunoregulatory lymphokines of T hybridomas from AIDS patients: constitutive and inducible suppressor factors. Science 1984;225: 66–69.

53 Hofmann B, Odum N, Jakobsen BK, Platz P, Ryder LP, Nielsen JO, Gerstoft J, Svejgaard A: Immunological studies in the acquired immunodeficiency syndrome. II. Active suppression or intrinsic defect – investigated by mixing AIDS cells with HLA-DR identical normal cells. Scand J Immunol 1986;23:669–678.

54 Siegel JP, Djeu JY, Stocks NI, Masur H, Gelmann EP, Quinnan GV: Sera from patients with the acquired immunodeficiency syndrome inhibit production of interleukin-2 by normal lymphocytes. J Clin Invest 1985;75:1957–1964.

55 Joly P, Guillon J-M, Mayaud C, Plata F, Theodorou I, Denis M, Delve P, Austran B: Cell-mediated suppression of HIV-specific cytotoxic T lymphocytes. J Immunol 1989;143:2193–2201.

56 Katz DH, Paul WE, Goidl EA, Benacerraf B: Carrier function in anti-hapten antibody responses. III. Stimulation of antibody synthesis and facilitation of hapten-specific secondary antibody responses by graft-versus-host reactions. J Exp Med 1971;133:169–186.

57 Margolick JB, Volkman DJ, Folks TM, Fauci AS: Amplification of HTLV-III/LAV infection by antigen-induced activation of T cells and direct suppression by virus lymphocyte blastogenic responses. J Immunol 1987;138:1719–1723.

58 Weinhold KJ, Lyerly HK, Stanley SD, Austin AA, Matthews TJ, Bolognesi DP: HIV-1 gp120-mediated immune suppression and lymphocyte destruction in the absence of viral infection. J Immunol 1989;142:3091–3097.

59 Diamond DC, Sleckman BP, Gregory T, Lasky LA, Greenstein JL, Burakoff SJ: Inhibition of CD4+ T cell function by the envelope glycoprotein gp120. J Immunol 1988;141:3715–3717.

60 Chirmule N, Kalyanaraman VS, Oyaizu N, Slade HB, Pahwa S: Inhibition of functional properties of tetanus antigen-specific T-cell clones by envelope glycoprotein gp120 of human immunodeficiency virus. Blood 1990;75:152–159.

61 Mittler RS, Hofmann MK: Synergism between gp120 and gp120-specific antibody in blocking human T-cell activation. Science 1989;245:1380–1382.

62 Golding H, Robey FA, Gates FT, Linder W, Beining PR, Hoffman T, Golding B: Identification of homologous regions in human immunodeficiency virus I gp41 and human MHC class II β1 domain. J Exp Med 1988;167:914–923.

63 Golding H, Shearer GM, Hillman K, Lucas P, Manischewitz J, Zajac RA, Clerici M, Gress RE, Boswell RN, Golding B: Common epitope in human immunodeficiency virus (HIV)-1 gp41 and HLA class II elicits immunosuppressive autoantibodies

capable of contributing to immune dysfunction in HIV-1-infected individuals. J Clin Invest 1989;83:1430–1435.

64 Viscidi RP, Mayur K, Lederman HM, Frankel RD: Inhibition of antigen-induced lymphocyte proliferation by *tat* protein from HIV-1. Science 1989;246:1606–1608.

65 Kekow J, Wachsman W, McCutchan JA, Cronin M, Carson DA, Lotz M: Transforming growth factor β and noncytopathic mechanisms of immunodeficiency in human immunodeficiency virus infection. Proc Natl Acad Sci USA 1990;87:8321–8325.

66 Volderbing P, Lagakos SW, Koch MA, Pettinelli C, Myers MW, Booth DK, Balfour HH, Reichaman RC, Bartlett JA, Hirsch MS, Murphy RL, Hardy WD, Soeiror-Fische MA, Bartlett JG, Merigan TC, Hyslop NE, Richman DD, Valentine FT, Corey L: Zidovudine in asymptomatic human immunodeficiency virus infection. A controlled trial in persons with fewer than 500 CD4-positive cells per cubic millimeter. N Engl J Med 1990;322:941–949.

67 Lambert JS, Seidlin M, Reichman RC, Plank CS, Laverty M, Morse GD, Knupp C, McLaren C, Pettinelli C, Valentine FT, Dolin R: 2′,3′-Dideoxyinosine (ddI) in patients with the acquired immunodeficiency syndrome or AIDS-related complex: A phase I trial. N Engl J Med 1990;322:1333–1340.

68 Cooley TP, Kunches LM, Saunders CA, Ritter JK, Perkins CJ, McLaren C, McCaffrey RP, Liebman HA: Once-daily administration of 2′,3′-dideoxyinosine (ddI) in patients with the acquired immunodeficiency syndrome or AIDS-related complex: Results of a phase I trial. N Engl J Med 1990;322:1340–1345.

69 Merrigan TC, Skowron G, Bozzette SA, Richman D, Uttamchandani R, Fischl M, Schooley R, Hirsch M, Soo W, Pettinelli C, Schaumberg H: The ddC study group of the AIDS clinical trials group: Circulating p24 antigen levels and responses to dideoxycytidine in human immunodeficiency virus (HIV) infections. A phase I and II study. Ann Intern Med 1989;110:189–194.

70 Yarchoan R, Pluda JM, Thomas RV, Mitsuya H, Brouwers P, Wyvill KM, Hartman N, Johns DG, Broder S: Long-term toxicity/activity profile of 2′3′-dideoxyinosine in AIDS or AIDS-related complex. Lancet 1990;336:526–529.

71 Poli G, Bressler P, Kinter A, Duh E, Justement JS, Stanley S, Fauci AS: Interleukin-6 induces human immunodeficiency virus expression in infected monocytic cells alone and in synergy with tumor necrosis factor alpha by transcriptional and post-transcriptional mechanisms. J Exp Med 1990;172:151–158.

72 Mosmann TR, Cherwinski H, Bond MW, Giedlin MA, Coffman RL: Two types of murine helper T cell clone. I. Definition according to profiles of lymphokine activities and secreted proteins. J Immunol 1986;136:2348–2357.

73 Cherwinski HM, Schumacher JH, Brown KD, Mosmann TR: Two types of mouse helper T cell clone. III. Further differences in lymphokine synthesis between Th1 and Th2 clones revealed by RNA hybridization, functionally monospecific bioassays, and monoclonal antibodies. J Exp Med 1987;166:1229–1244.

74 Mosmann TR, Coffman RL: Two types of mouse helper T cell clone: implications for immune regulation. Immunol Today 1987;8:223–227.

75 Zurawski G, Benedik M, Kamb BJ, Abrams JS, Zurawski SM, Lee FD: Activation of mouse T-helper cells induces abundant preproenkephalin mRNA synthesis. Science 1986;232:772–775.

76 Street NE, Schumacher JH, Fong TAT, Bass H, Fiorentino DF, Leverah JA, Mosmann TR: Heterogeneity of mouse helper T cells: evidence from bulk cultures

and limiting dilution cloning for precursors of Th1 and Th2 cells. J Immunol 1990;144:1629–1639.

77 Firestein GS, Roeder WD, Laxer JA, Townsend KS, Weaver CT, Horn JT, Linton J, Torbett BE, Glasebrook AL: A new murine CD4+ T cell subset with an unrestricted cytokine profile. J Immunol 1989;143:518–525.

78 Mosmann TR, Coffman RL: Heterogeneity of cytokine secretion patterns and functions of helper T cells. Adv Immunol 1989;46:111–147.

79 Ptak W, Janeway CA Jr, Flood PM: Immunoregulatory role of Ig isotypes. II. Activation of cells that block induction of contact sensitivity responses by antibodies of IgG2a and IgG2b isotypes. J Immunol 1988;141:765–773.

80 Ptak W, Flood PM, Janeway CA Jr, Marcinkiewicz J, Green DR: Immunoregulatory role of Ig isotypes. I. Induction by contrasuppressor T cells for contact sensitivity responses by antibodies of the IgM, IgG1 and IgG3 isotypes. J Immunol 1988; 141:756–764.

81 Fiorentino DF, Bond MW, Mosmann TR: Two types of mouse T helper cell. IV. TH2 clones secrete a factor that inhibits cytokine production by TH1 clones. J Exp Med 1989;170:2081–2095.

82 Mosmann TR, Moore KW: The role of IL-10 in cross-regulation of Th1 and Th2 responses; in Ash C, Gallagher RB (eds): Immunoparasitology Today. Cambridge/ UK, Elsevier Trends Journals, 1991, pp A49–A53.

83 DelPrete GF, DeCarli M, Mastromauro C, Biagiotti R, Macchia D, Falagiani P, Ricci M, Romagnani S: Purified protein derivative of *Mycobacterium tuberculosis* and excretory-secretory antigen(s) of *Toxocara canis* expand in vitro human T cells with stable and opposite (type 1 T helper or type 2 T helper) profile of cytokine production. J Clin Invest 1991;88:346–350.

84 Romagnani S: Human Th1 and Th2 subsets: doubt no more. Immunol Today 1991; 12:256–257.

85 Gaya A, DeLaCalle O, Yague J, Allsinet E, Fernandez MD, Romero M, Fabregat V, Martorell J, Vives J: IL-4 inhibits IL-2 synthesis and IL-2-induced up-regulation of IL-2 Rx but not IL-2Rβ chain in CD4+ human T cells. J Immunol 1991;146:4209–4214.

86 Yssel H, Shanafelt MC, Soderberg C, Schneider PV, Anzola J, Peltz G: *Borrelia burgdorferi* activates a T helper type 1-like T cell subset in Lyme arthritis. J Exp Med 1991;174:593–601.

87 Maggi E, Macchai D, Parrochi P, Mazzeti M, Ravina A, Milo D, Romagnani S: Reduced production of interleukin-2 and interferon-gamma and enhanced helper activity of IgG synthesis by cloned CD4+ T cells from patients with AIDS. Eur J Immunol 1987;17:1685–1690.

88 Lane CH, Masur H, Edgar LC, Whalen G, Rock AH, Fauci AS: Abnormalities of B cell activation and immunoregulation in patients with the acquired immunodeficiency syndrome. N Engl J Med 1983;309:453–455.

89 Chess Q, Daniels J, North E, Macris NT: Serum immunoglobulin elevations in the acquired immunodeficiency syndrome (AIDS): IgG, IgA, IgM, and IgD. Diagn Immunol 1984;2:148–153.

90 Kopelman RG, Zolla-Pazner S: Association of human immunodeficiency virus infection and autoimmune phenomena. Am J Med 1988;84:82–94.

91 Fling JA, Fischer JR, Boswell RN, Reid MN: The relationship of serum IgA concentration to human immunodeficiency (HIV) infection: A cross-sectional study

of HIV-seropositive individuals detected by screening in the United States Air Force. J Allergy Clin Immunol 1988;82:965–970.

92 Lucey DR, Zajac RA, Melcher GP, Butzin CA, Boswell RN: Serum IgE levels in 622 persons with human immunodeficiency virus infection: IgE elevation with marked depletion of CD4$^+$ T-cells. AIDS Res Hum Retroviruses 1990;6:427–429.

93 Nakajima K, Martinez-Maza O, Hirano T, Breen EC, Nishanian PG, Salazar-Gonzalez JF, Fahey JL, Kishimoto T: Induction of IL-6 (B cell stimulatory factor-2/IFN-β_2) production by HIV. J Immunol 1989;142:531–536.

94 Birx DL, Redfield RR, Tencer K, Fowler A, Burke DS, Tosato G: Induction of interleukin-6 during human immunodeficiency virus infection. Blood 1990;76: 2303–2310.

95 Ziegler JL, Beckstead JA, Volberding PA, Abrams DI, Levine AM, Lukes RJ, Gill PS, Burkes RL, Meyer PR, Metroka CE, Mouradian J, Moore A, Riggs SA, Butler JJ, Cabanillas FC, Hersh E, Newell GR, Laubenstein LJ, Knowles D, Odajnyk C, Raphael B, Koziner B, Urmacher C, Clarkson BD: Non-Hodgkin's lymphoma in 90 homosexual men: relation to generalized lymphadenopathy and the acquired immunodeficiency syndrome. N Engl J Med 1984;311:565–570.

96 Margalit H, Spouge JL, Cornette JL, Cease KB, DeLisi C, Berzofsky JA: Prediction of immunodominant helper T cell antigenic sites from the primary sequence. J Immunol 1987;138:2213–2219.

97 Cease KB, Margalit H, Cornette JL, Putney SD, Robey WG, Ouyang C, Streicher HZ, Fishchinger PJ, Gallo RC, Delisi C, Berzofsky JA: Helper T cell antigenic site identification in the acquired immunodeficiency syndrome virus gp120 envelope protein and induction of immunity in mice to the native protein using a 16-residue synthetic peptide. Proc Natl Acad Sci USA 1987;84:4249–4253.

98 Takahashi H, Cohen J, Hosmalin A, Cease KB, Houghten R, Cornette JL, DeLisi C, Moss B, Germain RN: An immunodominant epitope of the human immunodeficiency virus envelope glycoprotein gp160 recognized by class I major histocompatibility complex molecule-restricted murine cytotoxic T lymphocytes. Proc Natl Acad Sci USA 1988;85:3105–3109.

99 Hale PM, Cease KB, Houghten RA, Ouyang C, Putney S, Javaherian K, Margalit H, Cornette JL, Spouge JL, DeLisi C, Berzofsky JA: T cell multideterminant regions in the human immunodeficiency virus envelope: toward overcoming the problem of major histocompatibility complex restriction. Int Immunol 1989;1:409–418.

100 Clerici M, Stocks NI, Zajac RA, Boswell RN, Bernstein DC, Mann DL, Shearer DM, Berzofsky JA: Interleukin-2 production used to detect antigenic peptide recognition by T-helper lymphocytes from asymptomatic HIV-seropositive individuals. Nature 1989;339:383–385.

101 Clerici M, Tacket CO, Via CS, Lucey DR, Muluk SC, Zajac RA, Boswell RN, Berzofsky JA, Shearer GM: Immunization with subunit human immunodeficiency virus vaccine generates stronger T helper cell immunity than natural infection. Eur J Immunol 1991;21:1345–1349.

102 Clerici M, Berzofsky JA, Shearer GM, Tacket CO: Exposure to HIV-1 indicated by HIV-specific T helper cell responses before detection of infection by polymerase chain reaction and serum antibodies. J Infect Dis 1991;164:178–182.

103 Clerici M, Giorgi JV, Chen-Cheng C, Gudeman VK, Zack JA, Gupta P, Nishanian PG, Dudley JP, Berzofsky JA, Shearer GM: Specific anti-HIV-1 T cell immunity

in seronegative homosexuals with recent sexual exposure to seronegative partners. J Infect Dis 1991, in press.

104 Parish CR: The relationship between humoral and cell-mediated immunity. Trans Rev 1972;13:35–66.

105 Lagrange PH, Mackaness G, Miller T: Potentiation of T cell-mediated immunity by selective suppression of antibody formation with cyclophosphamide. J Exp Med 1974;139:1529–1539.

106 Bretscher PA: Regulation of the immune response by antigen. I. Specific T cells switch the in vivo response from a cell-mediated to a humoral mode. Cell Immunol 1983;81:345–352.

107 Dolin R, Graham BS, Greenberg SB, Tacket CO, Belshe RB, Midthun K, Clements ML, Gorse GJ, Horgan BW, Atmar RL, Karzon DT, Bonnez W, Fernie BF, Montefiori DC, Stablein DM, Smith GE, Koff WC, the NIAID AIDS Vaccine Clinical Trials Network: The safety and immunogenicity of a human immunodeficiency virus type 1 (HIV-1) recombinant gp160 candidate vaccine in humans. Ann Intern Med 1991;114:119–127.

108 Palca J: The sobering geography of AIDS. Science 1991;252:372–373.

109 Culliton BJ: AIDS against the rest of the world. Nature 1991;352:15.

110 Pearce EJ, Caspar P, Grzych JM, Lewis FA, Sher A: Down-regulation of Th1 cytokine production accompanies induction of Th2 responses by a parasitic Helminth, Schistosoma mansoni. J Exp Med 1991;173:159–166.

111 Locksley RM, Scott P: Helper T-cell subsets in mouse leishmaniasis: induction, expansion and effector function; in Ash C, Gallagher RB (eds): Immunoparasitology Today. Cambridge/UK, Elsevier Trends Journals, 1991, pp A59–A61.

112 Finkelman FD, Pearce EJ, Urban JF Jr, Sher A: Regulation and biological function of helminth-induced cytokine responses; in Ash C, Gallagher RB (eds): Immunoparasitology Today. Cambridge/UK, Elsevier Trends Journals, 1991, pp A62–A66.

113 Pearce EJ, Sher A: Minireview function dichotomy in the CD4$^+$ T cell response to Schistosoma mansoni. Exp Parasitol 1991;73:110–116.

114 Mosier DE, Yetter RA, Morse HC III: Retroviral induction of acute lymphoproliferative disease and profound immunosuppression in adult C57BL/6 mice. J Exp Med 1985;161:766–770.

115 Gazzinelli RT, Makino M, Chattopadhyay SK, Sher A, Hugin AW, Morse HC III: CD4$^+$ subset regulation in viral infection: preferential activation of Th2 cells during progression of metrovirus-induced immunodeficiency in mice. J Immunol 1991;148: 182–188.

116 Kullberg MC, Pearce EJ, Sher A, Berzofsky JA: Down-regulation of Th1-responses to myoglobin during Schistosoma mansoni infection (abstract). J Cell Biochem 1991; suppl 15F.

Gene M. Shearer, PhD, Experimental Immunology Branch, National Cancer Institute, National Institutes of Health, Bethesda, MD 20892 (USA)

Coffman RL (ed): Regulation and Functional Significance of T-Cell Subsets.
Chem Immunol. Basel, Karger, 1992, vol 54, pp 44–59

Evidence for Functional Subsets of CD4+ and CD8+ T Cells in Human Disease: Lymphokine Patterns in Leprosy

Padmini Salgame[a], Masahiro Yamamura[b], Barry R. Bloom[a], Robert L. Modlin[b]

[a] Howard Hughes Medical Institute, Albert Einstein College of Medicine, Bronx, N.Y., and [b] Division of Dermatology and Department of Microbiology and Immunology, UCLA School of Medicine, Los Angeles, Calif., USA

Introduction

Despite great progress in delineating antigens of a wide variety of viral, bacterial and parasitic pathogens, the precise immunological mechanisms required to engender resistance or protection, other than neutralizing antibodies, have been defined in very few infectious diseases of man. At present, the level of understanding of immune responses to intracellular pathogens is primarily by identification of T-lymphocyte subsets, as defined by cell surface antigen markers, found in lesions or responding to antigens in vitro. In animal models of infectious diseases, quite different outcomes of infection, e.g. protection or exacerbation, can be mediated by T cells of indistinguishable surface phenotype, but markedly different function. One of the key functional parameters determining the outcome of immune responses to such infectious agents is the nature of the lymphokines produced locally by immune T cells, yet the patterns of lymphokine production remain largely unknown for most infectious diseases of man.

Leprosy offers an attractive model for investigating the role of cytokines in resistance or susceptibility to infection for several reasons. Leprosy is a chronic infectious disease caused by *Mycobacterium leprae* that primarily affects skin, which means that the lesions are readily accessible to cellular and molecular analysis. Because leprosy is generally not life-threatening, dynamics of infiltrating cells can be investigated over time. Perhaps most impor-

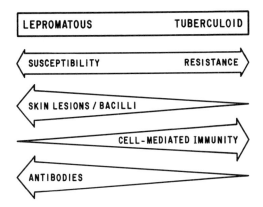

Fig. 1. The spectrum of leprosy.

tantly, it is a disease that presents as a clinical/immunologic spectrum [1]: At one end of the spectrum, patients with tuberculoid leprosy exemplify the resistant response that restricts the growth of the pathogen and the number of lesions, although frequently with some damage to nerves and tissues in the process. At the opposite end of this spectrum, patients with lepromatous leprosy represent extreme susceptibility to *M. leprae* infection. In lepromatous leprosy, disseminated skin lesions containing many viable *M. leprae* are characteristic.

It is widely agreed that T cells involved in cell-mediated immunity are pivotal in determining the outcome of infection with *M. leprae*, since skin test and lymphocyte reactivity are positive in tuberculoid patients, but negative in lepromatous patients. Yet there is an interesting paradox in that cell-mediated and humoral responses exhibit an inverse relationship (fig. 1). Anti-*M. leprae* antibodies are most elevated in patients with the lepromatous form of the disease, and are not thought to play a role in protection. We therefore sought to determine whether the resistance to *M. leprae* infection (associated with cell-mediated immunity) versus the susceptibility to *M. leprae* infection (associated with humoral responses) could be correlated with distinct cytokine patterns. The investigation was carried out at two levels: (1) cytokine patterns in leprosy lesions were detected by immunohistology, in situ hybridization and polymerase chain reaction (PCR), and (2) functionally distinct T cells from the leprosy lesions and blood were analyzed quantitatively for their lymphokine secretion patterns.

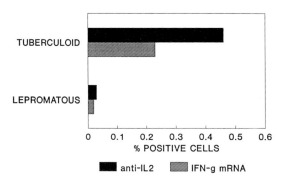

Fig. 2. IL-2- and IFN-γ-positive cells in leprosy lesions.

Detection of Cytokines in situ

IL-2 was initially detected in leprosy lesions by using monoclonal anti-IL-2 antibodies and immunoperoxidase techniques [2, 3]. An order of magnitude greater number of IL-2 containing cells were present in tuberculoid lesions as compared with lepromatous lesions (fig. 2). Additional immunohistochemical studies revealed that IFN-γ [4, 5], IL-1β [5] and TNF-α [5] positive cells were more numerous in tuberculoid than in lepromatous lesions.

The in situ hybridization allows for the detection of mRNA coding for cytokines in tissue sections of biopsy specimens. Since IFN-γ facilitates the intracellular killing of mycobacteria in vitro and in vivo [6, 7], leprosy skin biopsy specimens were examined for the presence of mRNA encoding this lymphokine [8]. Cells showing hybridization with an IFN-γ cDNA probe were more numerous in tuberculoid lesions in comparison to lepromatous lesions, and the percentages of positive cells were similar to those seen using the anti-IL-2 monoclonal antibody (fig. 2).

A role for TNF in protective immunity is implicated by the demonstration that this cytokine mediates the killing of mycobacteria [9] and granuloma formation [10]. Additionally, TNF mRNA and protein levels in lesions correlate with resistance to disease [5, 11, 12]. In tuberculoid lesions, 0.5% of the cells expressed TNF-α mRNA, whereas in lepromatous lesions, about 0.2% of the cells expressed TNF-α mRNA. Clearly, mycobacterial antigens induce TNF release and TNF can be detected in the serum and lesions of patients with mycobacterial disease. The pattern of TNF production and distribution in patients with mycobacterial disease, and the immune effects of TNF, indicate that TNF may also mediate pathologic manifestations of mycobacterial disease, including fever and tissue necrosis.

Detection of Cytokine mRNA in Lesions by PCR Amplification

To more fully probe patterns of cytokines in lesions at the extremes of the spectrum of leprosy, we used the following strategy [12]: (1) to extract total RNA from biopsy specimens; (2) reverse transcribe polyadenylated mRNAs to obtain lymphokine cDNAs, and (3) detect those cDNAs with high sensitivity by the PCR using cytokine-specific primers. Since PCR is recognized to be a semiquantitative technique at best, to provide meaningful comparisons between different individuals and forms of the disease the cDNAs were normalized, initially relative to β-actin mRNA which is found in all cells.

The striking observation was that clear differences in cytokine PCR profiles were observed between tuberculoid and lepromatous lesions, and those differences were consistent across lesions of all patients. Messenger RNAs encoding cytokines predominantly produced by macrophages, i.e., IL-1β, TNF-α, GM-CSF, TGF-β$_1$ and IL-6, were more abundant in the tuberculoid than the lepromatous lesions studied. Furthermore, cytokine mRNAs for IL-2, IFN-γ, and lymphotoxin produced by T cells were abundant in tuberculoid lesions, but virtually absent in lepromatous lesions. In contrast, IL-4 and to a lesser extent IL-10 mRNA were detected at higher levels in lepromatous than tuberculoid lesions. IL-3 and IL-5 were weak or absent in both lesions, IL-8 was present in both.

Where relatively subtle differences in levels of some lymphokine mRNAs were seen, the discriminatory capacity of the PCR results was enhanced by normalizing the PCR comparisons to specific CD3δ chain mRNA found only in T cells. The dichotomy between tuberculoid and lepromatous lesions observed was even more striking: IL-2 and IFN-γ mRNA were markedly higher in tuberculoid lesions, while IL-4, IL-5 and IL-10 were characteristic of lepromatous lesions (fig. 3).

There are few data available in human infection on the nature of the lymphokines and cytokines involved in engendering resistance or tissue damage. In murine models of intracellular pathogens, such as *Leishmania major*, *Schistosoma mansoni* and *Trichinella spiralis*, resistant versus susceptible immune responses appear to be regulated by two CD4$^+$ T-helper subpopulations which differ in their effector functions and cytokine patterns

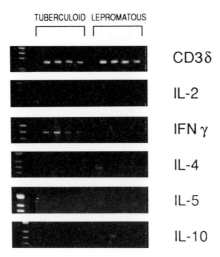

Fig. 3. PCR analysis of cytokine mRNAs from leprosy lesions.

[13–15]. T cells that produce IL-2 and IFN-γ, termed T_H1 cells [16], preferentially activate macrophages to kill or inhibit the growth of the pathogen, resulting in mild or self-curing disease. In contrast, T cells producing IL-4 and IL-5, termed T_H2 cells, augment humoral responses and inhibit cell-mediated immune responses, resulting in fulminant infection.

Similar functional populations have been elegantly demonstrated in man in response to different antigens [17–19]. In addition, a particular subset, the T_H2 population, has been associated with human diseases of allergic etiology [20], and *Loa loa* infection [21, 22]. Nevertheless, it had not been possible to define comparable T_H1 and T_H2 responses in man in response to a *single* pathogen. The lymphokine patterns that were detected in the leprosy lesions are remarkably similar to the T_H1 and T_H2 patterns of murine CD4 cells. T_H1-like lymphokine mRNAs were abundant in tuberculoid lesions, which are basically self-healing and characterized by resistance to growth of *M. leprae*. In marked contrast, T_H2-like lymphokine mRNAs were abundant in lepromatous lesions, correlating with immunologic unresponsiveness to *M. leprae*. These results provide evidence that such patterns regulate the human immune response to a given pathogen.

The abundance of IL-2, IFN-γ, lymphotoxin TNF-α and GM-CSF in tuberculoid lesions is likely to contribute to the resistant state of immunity in these patients. IFN-γ is well known to enhance production of reactive oxygen and nitrogen intermediates [23] by macrophages and stimulate them to kill or restrict the growth of intracellular pathogens [24]. IFN-γ also augments expression of HLA-DR and ICAM-1 which facilitates T-cell-accessory cell interaction [25]. IL-2 may contribute to the host defense by inducing the clonal expansion of immune activated lymphokine-producing T cells and augments the production of IFN-γ [26]. In fact, cutaneous administration of either IL-2 or IFN-γ to lepromatous patients results in some clearance of bacilli from lesions [7, 27].

In contrast to the set of cytokine mRNAs present in tuberculoid lesions that might be involved in cell-mediated immunity and inflammation, those found to be increased in lepromatous lesions might be expected to contribute to the immune unresponsiveness and failure of macrophage activation in these individuals. IL-4 was increased in lepromatous lesions compared with tuberculoid lesions, and is known to inhibit IFN-γ production [28] as well as IFN-γ-mediated production of reactive oxygen intermediates such as hydrogen peroxide by macrophages [29]. Additionally, IL-4 has been reported to inhibit IL-2 activation of T cells through IL-2R expression [30] and inhibit production of TNF-α and IL-1β by macrophages [31]. The IL-4 and IL-5 in lepromatous lesions could lead to similar enhancement of bacterial growth and even contribute to the elevated levels of anti-*M. leprae* Ig found in lepromatous patients [32, 33]. The presence of IL-10 in lepromatous lesions, another cytokine that can down-regulate lymphokine production by CD4 T cells, in the relative absence of IFN-γ or IL-2, suggests a possible role for this cytokine in the specific immunological unresponsiveness to *M. leprae* antigens [34, 35].

Lymphokine Profiles of CD4 and CD8 T Clones

The lymphocyte profiles of skin lesions of leprosy provided a valuable window into the patterns of cell-mediated immune responses throughout the spectrum of leprosy. Earlier immunoperoxidase and immunofluorescence studies had indicated differences in the CD4:CD8 ratio (T-helper:T-suppressor) at the poles of the leprosy spectrum (fig. 4) [2, 3, 36–48]. The data from these studies indicate that in tuberculoid leprosy lesions the CD4 population predominated with a CD4:CD8 ratio of 1.9:1, whereas in the lepromatous lesions the CD8 population predominated with a CD4:CD8 ratio of 0.6:1. Furthermore, CD4 cells in tuberculoid lesions mark as T-memory cells

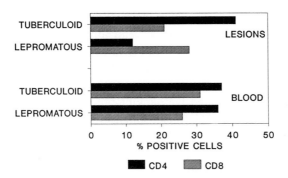

Fig. 4. CD4 and CD8 populations in lesions and blood of leprosy patients.

(CD45R0+) and those in lepromatous lesions express the T-naive phenotype (CD45R+) [49]. The majority of CD8+ cells are CD28-, indicating that they are of the T-suppressor phenotype [49]. CD8+ cells of the T-cytotoxic phenotype (CD28+) predominate in the tuberculoid lesions. Of CD4 cells cultured directly from tuberculoid lesions, 1 in 60 react to *M. leprae* antigens [49]. CD8 cells from lepromatous lesions fail to proliferate to antigen, but can be activated by *M. leprae* to suppress proliferative responses by CD4 cells [50, 51]. In order to gain insight into the molecular and cellular basis of T-cell-mediated protection and suppression, as well as to correlate the lymphokine patterns in lesions with functional subsets of CD4 and CD8 T cells, we examined the profile of lymphokines produced by T-cell clones derived from patients across the spectrum of leprosy [52]. The patterns of lymphokine secretion of these clones were compared to the lymphokine profiles of cells specific for other antigens bearing the same surface phenotype, but differing in function.

A panel of *M. leprae*-specific T_S and T_H clones from lesions and peripheral blood of leprosy patients and lepromin-positive familial contacts was established. For purposes of comparison, eight CD4 clones specific for tetanus toxoid and one for influenza hemagglutinin (Flu), and six CD8 MHC class I restricted T_C cells specific for HLA-B27 were studied. The T-cell clones were stimulated via the T-cell receptor–CD3 complex using anti-CD3 monoclonal antibodies, and the supernatants were harvested at 18 h and tested by immunoenzymatic assay for IL-2, IL-4, IL-5, IL-6, IL-10, IFN-γ, GM-CSF and TNF-α (fig. 5). All (13/13) of the *M. leprae*-specific CD4+ clones obtained from antigen-responsive patients or contacts and the Flu clone produced IFN-γ, TNF-α and GM-CSF, but little or no IL-4 or IL-5, similar to the

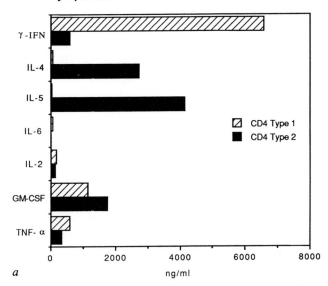

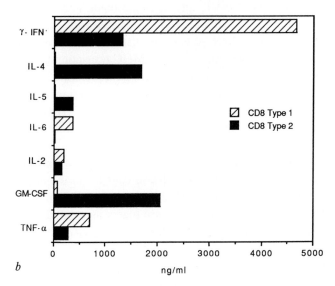

Fig. 5. Cytokine patterns of *(a)* CD4[+] and *(b)* CD8[+] T-cell clones stimulated by immobilized anti-CD3.

pattern of lymphokines characteristic of murine T_H1 cells. These clones also lacked helper activity for antibody formation, i.e. inducing IgM production by the SKW6.4 human B-cell line. We have therefore designated them as 'type 1 CD4 cells'. The CD4 tetanus toxoid clones produced low levels of IFN-γ and IL-2 but, with one exception, high levels of IL-4, IL-5 and GM-CSF, a pattern similar to T_H2 cells, and were designated as 'type 2 CD4 cells'. Two CD4$^+$ tetanus-specific clones that produced large amounts of IL-4 were found to have B-cell helper activity. Thus, functionally different CD4 cells from strongly antigen-reactive donors can be subtyped into two groups, similar to the murine counterpart, based on lymphokine patterns.

Of the six CD8$^+$ alloreactive T_C clones tested, five secreted IFN-γ, IL-6, TNF-α and IL-10, but made no detectable IL-4 or IL-5. Interestingly, the pattern of lymphokine production by the CD8 T_S clones was characterized by high levels of IL-4 and GM-CSF but little or no IL-6 and IL-10. Although the CD8$^+$ T_S clones produced high levels of IL-4, only very low levels of IL-5 were produced, unlike the type 2 CD4 clones. It was surprising that the CD8$^+$ T_S clones did not make any IL-10 which has been reported to be an immunosuppressive molecule [53, 54] and its mRNA was also found in the lesions of lepromatous patients. It may well be that much of the IL-10 in the lesion is contributed by macrophages and/or CD4$^+$ type 2 cells. The T_S clones also made low levels of IFN-γ. Both the T_C and T_S clones made TNF-α. With the possible exception of the CD28 surface marker associated with T_C but not T_S cells [55], it has not previously been possible to discriminate between CD8$^+$ T_C and T_S cells other than by their function and MHC restriction for antigen (CD8$^+$ T_C are generally MHC class I restricted, and the T_S cells are MHC class II restricted) [50, 56]. Based on patterns of lymphokine secretion, particularly of IL-4, IL-6 and IL-10, the data suggest that the human CD8 population can also be divided into two functional subsets, that we designate as 'type 1' and 'type 2 CD8 cells'.

Two significant differences in the lymphokine pattern of mouse and human clones emerged. Among the CD4 cells, in contrast to the mouse, all the human clones we examined made IL-2 and most made some IFN-γ. Second, although IL-4 and IL-5 production are highly associated in mouse T_H2 cells, IL-5 was not produced in abundance by CD8 T_S clones which produced high levels of IL-4.

The striking finding of this analysis was the central role of IL-4 production in permitting discrimination between the functional subsets of both CD4 and CD8 T-cell subpopulations. IL-4, initially described as a B-cell-stimulating factor, has been shown to mediate multiple biologic functions in a variety of cell types. It is best known for its necessary role in activation and immunoglobulin class switching of B cells [57]. In addition, IL-4 stimulates murine T_H2 proliferation in the presence of IL-1 [58]. IL-4 also has a negative immunoregulatory role, since it: (1) blocks IL-2-dependent proliferation of human T cells by down-regulation of IL-2 receptors [30]; (2) abrogates both the IFN-γ-mediated activation of monocytes and their antileishmanial activity [29]; (3) down-regulates CD14 expression on monocytes and production of IL-1 and TNF-α [31, 59], and (4) blocks macrophage nitric oxide generation necessary for killing intracellular parasites [60]. There is growing evidence that IL-4 and IFN-γ mutually inhibit their respective functions. IFN-γ is reported to inhibit the proliferation of T_H2 cells [61] and IL-4 inhibits IFN-γ production by human lymphocytes [28]. The in vivo relevance of these findings has been best demonstrated in the murine leishmanias model. In *L. major* infections, healing in C57/BL mice is accompanied by an increase in IFN-γ production by leishmania-antigen-specific cells and in the susceptible BALB/c mice, there is an increase in IL-4 production [13, 62–65]. Based on these observations, Locksley and co-workers [64] injected BALB/c mice with neutralizing anti-IL-4 antibody, and observed an attenuation of disease in 100% and cure in 85% of the animals [64]. Recognition of different antigens of leishmania may favor T_H1 or T_H2 responses [62]. Specifically, protective T_H1 were preferentially generated by living and not killed parasites. In murine schistosome infection, lymphocytes from animals immunized with larval antigens made IFN-γ and IL-2, while those from egg infected animals made IL-4 and IL-5 [14]. The T_H2 response in infected animals was induced mainly by the egg antigens and the protective T_H1 response was made to the larval antigens. Initiation of egg-laying in infected animals making IFN-γ, resulted in down-regulation of protective T_H1 cytokine production. In congenic mice resistant and susceptible to *T. spiralis* infection, resistance was associated with IFN-γ and IgG2a antibodies, and susceptibility with IL-4, IgE and IgG1 antibodies [15]. The reciprocal modulation of activities by IL-4 and IFN-γ has been documented in man as well. In filariasis, for example, IgE production to filarial antigens is mediated by IL-4 and down-regulated by IFN-γ [21].

It is a well-known immunological generalization that a dichotomy exists between antibodies to a pathogen and cell-mediated immunity. As a case inpoint, lepromatous leprosy patients, lacking CMI, have significantly higher levels of anti-*M. leprae* antibodies than tuberculoid patients. These data and the results of murine infection models can best be reconciled by appreciating that IL-4 has the capacity not only to enhance antibody formation, but also to depress multiple components of cell-mediated immunity required for protection. However, it should be noted that there may be significant contributions to the outcome of infection by other lymphokines and cytokines detected in lesions and supernatants of *M. leprae*-reactive clones. It is unclear whether cytokines act individually or additively. More likely it is the matrix of lymphokines and cytokines that determine the ultimate biological response at the locus of infection, protective immunity and/or immunopathology.

A number of intriguing questions remain, the most compelling perhaps being what determines the particular functional T-cell subset that will be principally engaged by antigen in a particular context. There is a suggestion in the parasitic models that the nature of the antigens and/or epitopes recognized by these T-cell subsets may be important [62]. The nature of the antigen-presenting cell may be a determining factor. For example, in the mouse, macrophages may preferentially present antigen to T_H1 cells, but B cells present to T_H2 cells [66]. The MHC-restricting element may be another variable that determines the nature of the T-cell subset response. In the case of leprosy, the majority of CD4-responding cells from tuberculoid patients are restricted by HLA-DR [67], but all the CD8 T_S clones studied by us have been restricted by HLA-DQ [56]. The MHC restrictions favoring antigen presentation to type 2 CD4 cells has not been defined in any system, and it would be of interest to learn if it is biased to HLA-DQ as well. Other events which may determine the type of T-cell response include antigen dose, the interplay of accessory molecules and the so-called 'second signals' which may be required for activation.

The studies of leprosy clearly demonstrate that local production of the T_H1 and T_H2 lymphokines can be associated with resistant versus susceptible responses to infection in man, and that functional subsets of human CD4+ and CD8+ populations can be assigned on the basis of these lymphokine patterns. Moreover, the data from lesions indicates that lymphokine profiles may to a large degree be determinative of the outcome of specific infectious challenges to host defenses of man.

Acknowledgments

Supported by grants from the National Institutes of Health (AI 22553, AR 40312, AI 40312, AI 07118 and AI 20111 and AI 26491), the UNDP/World Bank/World Health

Organization Special Programme for Research and Training in Tropical Diseases (IMMLEP and THELEP), the Heiser Trust and the Sasakawa Memorial Health Foundation.

References

1 Ridley DS, Jopling WH: Classification of leprosy according to immunity. A five-group system. Int J Lepr 1966;34:255–273.

2 Modlin RL, Hofman FM, Horwitz DA, Husmann LA, Gillis S, Taylor CR, Rea TH: In situ identification of cells in human leprosy granulomas with monoclonal antibodies to interleukin-2 and its receptor. J Immunol 1984;132:3085–3090.

3 Longley J, Haregewoin A, Yemaneberhan T, van Diepen TW, Nsibami J, Knowles D, Smith KA, Godal T: In vivo responses to *Mycobacterium leprae*: antigen presentation, interleukin-2 production, and immune cell phenotypes in naturally occurring leprosy lesions. Int J Lepr 1985;53:385–394.

4 Volc-Platzer B, Stemberger H, Luger T, Radaszkiewicz T, Wiedermann G: Defective intralesional interferon-gamma activity in lepromatous leprosy. Clin Exp Immunol 1988;71:235–240.

5 Arnoldi J, Gerdes J, Flad H-D: Immunohistologic assessment of cytokine production of infiltrating cells in various forms of leprosy. Am J Pathol 1990;137:749–753.

6 Rook GAW, Steele J, Fraher L, Barker S, Karmali R, O'Riordan J: Vitamin D_3, gamma interferon, and control of proliferation of *Mycobacterium tuberculosis* by human monocytes. Immunology 1986;57:159–163.

7 Nathan CF, Kaplan G, Levis WR, Nusrat A, Witmer MD, Sherwin SA, Job CK, Horowitz CR, Steinman RM, Cohn ZA: Local and systemic effects of intradermal recombinant interferon-gamma in patients with lepromatous leprosy. N Engl J Med 1986;315:6–15.

8 Cooper CL, Mueller C, Sinchaisri T-A, Pirmez C, Chan J, Kaplan G, Young SMM, Weissman IL, Bloom BR, Rea TH, Modlin RL: Analysis of naturally occurring delayed-type hypersensitivity reactions in leprosy by in situ hybridization. J Exp Med 1989;169:1565–1581.

9 Bermudez LEM, Young LS: Tumor necrosis factor, alone or in combination with IL-2, but not IFN-gamma, is associated with macrophage killing of *Mycobacterium avium* complex. J Immunol 1988;140:3006–3013.

10 Kindler V, Sappino A-P, Grau GE, Piguet P-F, Vassalli P: The inducing role of tumor necrosis factor in the development of bactericidal granulomas during BCG infection. Cell 1989;56:731–740.

11 Sullivan L, Sano S, Pirmez C, Salgame P, Mueller C, Hofman F, Uyemura K, Rea TH, Bloom BR, Modlin RL: Expression of adhesion molecules in leprosy lesions. Infect Immun 1991;59:4154–4160.

12 Yamamura M, Uyemura K, Deans RJ, Weinberg K, Rea TH, Bloom BR, Modlin RL: Defining protective responses to infectious pathogens: cytokine profiles in leprosy lesions. Science 1991;254:277–279.

13 Heinzel FP, Sadick MD, Holaday BJ, Coffman RL, Locksley RM: Reciprocal expression of interferon-gamma or interleukin-4 during the resolution or progression of murine leishmaniasis. Evidence for expansion of distinct helper T cell subsets. J Exp Med 1989;169:59–72.

14 Pearce EJ, Caspar P, Grzych J-M, Lewis FA, Sher A: Down-regulation of Th1 cytokine production accompanies induction of Th2 responses by a parasitic helminth, *Schistosoma mansoni*. J Exp Med 1991;173:159–166.

15 Pond L, Wassom DL, Hayes CE: Evidence for differential induction of helper T cell subsets during *Trichinella spiralis* infection. J Immunol 1989;143:4232–4237.

16 Mosmann TR, Cherwinski H, Bond MW, Giedlin MA, Coffman RL: Two types of murine helper T cell clone. I. Definition according to profiles of lymphokine activities and secreted proteins. J Immunol 1986;136:2348–2357.

17 Maggi E, Biswas P, Del Prete G, Parronchi P, Macchia D, Simonelli C, Emmi L, De Carli M, Tiri A, Ricci M, Romagnani S: Accumulation of Th-2-like helper T cells in the conjunctiva of patients with vernal conjunctivitis. J Immunol 1991;146:1169–1174.

18 Parronchi P, Macchia D, Piccinni MP, Biswas P, Simonelli C, Maggi E, Ricci M, Ansari AA, Romagnani S: Allergen- and bacterial antigen-specific T-cell clones established from atopic donors show a different profile of cytokine production. Proc Natl Acad Sci USA 1991;88:4538–4542.

19 Del Prete GF, De Carli M, Mastromauro C, Biagiotti R, Macchia D, Falagiani P, Ricci M, Romagnani S: Purified protein derivative of *Mycobacterium tuberculosis* and excretory-secretory antigen(s) of *Toxocara canis* expand in vitro human T cells with stable and opposite (type 1 T-helper or type 2 T-helper) profile of cytokine production. J Clin Invest 1991;88:346–350.

20 Wierenga EA, Snoek M, de Groot C, Chretien I, Bos JD, Jansen HM, Kapsenberg ML: Evidence for compartmentalization of functional subsets of CD4$^+$ T lymphocytes in atopic patients. J Immunol 1990;144:4651–4656.

21 King CL, Ottesen EA, Nutman TB: Cytokine regulation of antigen-driven immunoglobulin production in filarial parasite infections in humans. J Clin Invest 1990;85: 1810–1815.

22 Limaye AP, Abrams JS, Silver JE, Ottensen EA, Nutman TB: Regulation of parasite-induced eosinophilia: selectively increased interleukin-5 production in helminth-infected patients. J Exp Med 1990;172:399–402.

23 Nathan CF, Murray HW, Wiebe ME, Rubin BY: Identification of interferon-gamma as the lymphokine that activates human macrophage oxidative metabolism and antimicrobial activity. J Exp Med 1983;158:670–689.

24 Murray HW, Rubin BY, Rothermel CP: Killing of intracellular *L. donovani* by lymphokine-stimulated human mononuclear phagocytes. Evidence that interferon is the activating lymphokine. J Clin Invest 1983;72:1506–1510.

25 Dustin ML, Singer KH, Tuck DT, Springer TA: Adhesion of T lymphoblasts to epidermal keratinocytes is regulated by interferon-gamma and is mediated by intercellular adhesion molecule 1 (ICAM-1). J Exp Med 1988;167:1323–1340.

26 Kasahara T, Hooks JJ, Dougherty SF, Oppenheim JJ: Interleukin-2-mediated immune interferon (IFN-gamma) production by human T cells and T cell subsets. J Immunol 1983;130:1784–1789.

27 Kaplan G, Kiessling R, Teklemariam S, Hancock G, Sheftel G, Job CK, Converse P, Ottenhoff THM, Becx-Bleumink M, Dietz M, Cohn ZA: The reconstitution of cell-mediated immunity in the cutaneous lesions of lepromatous leprosy by recombinant interleukin-2. J Exp Med 1989;169:893–907.

28 Peleman R, Wu J, Fargeas C, Delespesse G: Recombinant interleukin-4 suppresses the production of interferon-gamma by human mononuclear cells. J Exp Med 1989; 170:1751–1756.

29 Lehn M, Weiser WY, Engelhorn S, Gillis S, Remold HG: IL-4 inhibits H_2O_2 production and antileishmanial capacity of human cultured monocytes mediated by IFN-gamma. J Immunol 1989;143:3020–3024.

30 Martinez OM, Gibbons RS, Garovoy MR, Aronson FR: IL-4 inhibits IL-2 receptor expression and IL-2-dependent proliferation of human T cells. J Immunol 1990;144:2211–2215.

31 Hart PH, Vitti GF, Burgess DR, Whitty GA, Piccoli DS, Hamilton JA: Potential anti-inflammatory effects of interleukin-4: Suppression of human monocyte tumor necrosis factor alpha, interleukin-1, prostaglandin E_2. Proc Natl Acad Sci USA 1989; 86:3803–3807.

32 Yokota T, Otsuka T, Mosmann T, Banchereau J, DeFrance T, Blanchard D, De Vries JE, Lee F, Arai K: Isolation and characterization of a human interleukin cDNA clone, homologous to mouse B-cell stimulatory factor 1, that expresses B-cell- and T-cell-stimulating activities. Proc Natl Acad Sci USA 1986;83:5894–5898.

33 Azuma C, Tanabe T, Konishi M, Kinashi T, Noma T, Matsuda F, Yaoita Y, Takatsu K, Hammarstrom L, Edvard Smith CI, Severinson E, Honjo T: Cloning of cDNA for human T-cell replacing factor (interleukin-5) and comparison with the murine homologue. Nucleic Acids Res 1986;14:9149–9158.

34 Moore KW, Vieira P, Fiorentino DF, Trounstine ML, Khan TA, Mosmann TR: Homology of cytokine synthesis inhibitory factor (IL-10) to the Epstein-Barr virus gene BCRFI. Science 1990;248:1230–1234.

35 Vieira P, de Waal-Malefyt R, Dang M-N, Johnson KE, Kastelein R, Fiorentino DF, de Vries JE, Roncarolo M-G, Mosmann TR, Moore KW: Isolation and expression of human cytokine synthesis inhibitory factor cDNA clones: homology to Epstein-Barr virus open reading frame BCRFI. Proc Natl Acad Sci USA 1991;88:1172– 1176.

36 Modlin RL, Hofman FM, Taylor CR, Rea TH: In situ characterization of T lymphocyte subsets in leprosy granulomas (letter). Int J Lepr 1982;50:361–362.

37 Van Voorhis WC, Kaplan G, Sarno EN, Horwitz MA, Steinman RM, Levis WR, Nogueira N, Hair LS, Gattass CR, Arrick BA, Cohn ZA: The cutaneous infiltrates of leprosy: cellular characteristics and the predominant T-cell phenotypes. N Engl J Med 1982;307:1593–1597.

38 Modlin RL, Hofman FM, Taylor CR, Rea TH: T lymphocyte subsets in the skin lesions of patients with leprosy. J Am Acad Dermatol 1983;8:182–189.

39 Salgame PR, Mahadevan PR, Antia NH: Mechanism of immunosuppression in leprosy: presence of suppressor factor(s) from macrophages of lepromatous patients. Infect Immun 1983;40:1119–1126.

40 Narayanan RB, Bhutani LK, Sharma AK, Nath I: T cell subsets in leprosy lesions: in situ characterization using monoclonal antibodies. Clin Exp Immunol 1983;51:421–429.

41 Modlin RL, Gebhard JF, Taylor CR, Rea TH: In situ characterization of T lymphocyte subsets in the reactional states of leprosy. Clin Exp Immunol 1983;53: 17–24.

42 Kato H, Sanada K, Koseki M, Ozawa T: Identification of lymphocyte subpopulations in cutaneous lesions of leprosy. Nippon Rai Gakkai Zasshi 1983;52:126–132.

43 Wallach D, Flageul B, Bach MA, Cottenot F: The cellular content of dermal leprous granulomas: an immunohistological approach. Int J Lepr 1984;52:318–326.

44 Sarno EN, Kaplan G, Alvaranga F, Nogueira N, Porto JA, Cohn ZA: Effect of treatment on the cellular composition of cutaneous lesions in leprosy patients. Int J Lepr 1984;52:496–500.

45 Narayanan RB, Laal S, Sharma AK, Bhutani LK, Nath I: Differences in predominant T cell phenotypes and distribution pattern in reactional lesions of tuberculoid and lepromatous leprosy. Clin Exp Immunol 1984;55:623–628.
46 Modlin RL, Bakke AC, Vaccaro SA, Horwitz DA, Taylor CR, Rea TH: Tissue and blood T-lymphocyte subpopulations in erythema nodosum leprosum. Arch Dermatol 1985;121:216–219.
47 Modlin RL, Mehra V, Jordan R, Bloom BR, Rea TH: In situ and in vitro characterization of the cellular immune response in erythema nodosum leprosum. J Immunol 1986;136:883–886.
48 Nilsen R, Mshana RN, Negesse Y, Menigistu G, Kana B: Immunohistochemical studies of leprous neuritis. Lepr Rev 1986;57(suppl 2):177–187.
49 Modlin RL, Melancon-Kaplan J, Young SMM, Pirmez C, Kino H, Convit J, Rea TH, Bloom BR: Learning from lesions: Patterns of tissue inflammation in leprosy. Proc Natl Acad Sci USA 1988;85:1213–1217.
50 Modlin RL, Kato H, Mehra V, Nelson EE, Xue-dong F, Rea TH, Pattengale PK, Bloom BR: Genetically restricted suppressor T-cell clones derived from lepromatous leprosy lesions. Nature 1986;322:459–461.
51 Modlin RL, Mehra V, Wong L, Fujimiya Y, Chang W-C, Horwitz DA, Bloom BR, Rea TH, Pattengale PK: Suppressor T lymphocytes from lepromatous leprosy skin lesions. J Immunol 1986;137:2831–2834.
52 Salgame P, Abrams J, Clayberger C, Goldstein H, Modlin RL, Bloom BR: Differing lymphokine profiles of functional subsets of human CD4 and CD8 T cell clones. Science 1991;254:279–282.
53 Fiorentino DF, Bond MW, Mosmann TR: Two types of mouse T helper cell. IV. Th2 clones secrete a factor that inhibits cytokine production by Th1 clones. J Exp Med 1989;170:2081–2095.
54 Fiorentino DF, Zlotnik A, Vieira P, Mosmann TR, Howard M, Moore KW, OGarra A: IL-10 acts on the antigen-presenting cell to inhibit cytokine production by Th1 cells. J Immunol 1991;146:3444–3451.
55 Yamada H, Martin PJ, Bean MA, Braun MP, Beatty PG, Sadamoto K, Hansen JA: Monoclonal antibody 9.3 and anti-CD11 antibodies define reciprocal subsets of lymphocytes. Eur J Immunol 1985;15:1164–1168.
56 Salgame P, Convit J, Bloom BR: Immunological suppression by human CD8+ T cells is receptor dependent and HLA-DQ restricted. Proc Natl Acad Sci USA 1991;88:2598–2602.
57 Finkelman FD, Holmes J, Katona IM, Urban JF, Beckmann MP, Park LS, Schooley KA, Coffman RL, Mosmann TR, Paul WE: Lymphokine control of in vivo immunoglobulin isotype selection. Annu Rev Immunol 1990;8:303–333.
58 Fernandez Botran R, Sanders VM, Mosmann TR, Vitetta ES: Lymphokine-mediated regulation of the proliferative response of clones of T-helper 1 and T-helper 2 cells. J Exp Med 1988;168:543–558.
59 Hart PH, Cooper RL, Finlay Jones JJ: IL-4 suppresses IL-1-beta, TNF-alpha and PGE$_2$ production by human peritoneal macrophages. Immunology 1991;72:344–349.
60 Liew FY, Cox FE: Nonspecific defence mechanism: the role of nitric oxide. Immunol Today 1991;12:A17–A21.
61 Gajewski TF, Fitch FW: Anti-proliferative effect of IFN-gamma in immune regulation. I. IFN-gamma inhibits the proliferation of Th2 but not Th1 murine helper T lymphocyte clones. J Immunol 1988;140:4245–4252.

62 Scott P, Natovitz P, Coffman RL, Pearce E, Sher A: Immunoregulation of cutaneous leishmaniasis. T cell lines that transfer protective immunity or exacerbation belong to different T helper subsets and respond to distinct parasite antigens. J Exp Med 1988;168:1675–1684.

63 Liew FY, Millott S, Li Y, Lelchuk R, Chan WL, Ziltener H: Macrophage activation by interferon-gamma from host-protective T cells is inhibited by interleukin IL-3 and IL-4 produced by disease-promoting T cells in leishmaniasis. Eur J Immunol 1989;19:1227–1232.

64 Sadick MD, Heinzel FP, Holaday BJ, Pu RT, Dawkins RS, Locksley RM: Cure of murine leishmaniasis with anti-interleukin-4 monoclonal antibody. Evidence for a T-cell-dependent, interferon-gamma-independent mechanism. J Exp Med 1990;171: 115–127.

65 Boom WH, Liebster L, Abbas AK, Titus RG: Patterns of cytokine secretion in murine leishmaniasis: correlation with disease progression or resolution. Infect Immun 1990;58:3863–3870.

66 Gajewski TF, Pinnas M, Wong T, Fitch FW: Murine Th1 and Th2 clones proliferate optimally in response to distinct antigen-presenting cell populations. J Immunol 1991;146:1750–1758.

67 Ottenhoff TH, Neuteboom S, Elfrink DG, de Vries RR: Molecular localization and polymorphism of HLA class II restriction determinants defined by *Mycobacterium leprae*-reactive helper T cell clones from leprosy patients. J Exp Med 1986;164: 1923–1939.

Robert L. Modlin, MD, UCLA Division of Dermatology, 52-121 CHS,
10833 Le Conte Avenue, Los Angeles, CA 90024-1750 (USA)

Coffman RL (ed): Regulation and Functional Significance of T-Cell Subsets.
Chem Immunol. Basel, Karger, 1992, vol 54, pp 60–71

T-Cell Subsets in Experimental Allergic Encephalomyelitis: Interactions between the Immune and Neuroendocrine Systems in the Regulation of Their Activity

Don Mason

MRC Cellular Immunology Unit, Sir William Dunn School of Pathology,
Oxford, UK

Introduction

Although the pleiotropic properties of the cytokines involved in the immune response ensure that the immune system is a highly interconnected one, it is evident that it does not function in isolation. In particular, it is becoming increasingly apparent that the neuroendocrine system can have a potent regulatory effect on an immune reaction. It has been suggested that this regulation has evolved to prevent an immune response to a foreign organism from damaging the host that produces it [1]. The recent finding that cytokines can stimulate the hypothalamic-pituitary-adrenal system with the release of glucocorticoids has provided support for this view [2, 3].

In parallel with these developments in establishing links between the immune and neuroendocrine systems there has been a notable increase in our knowledge of T-cell heterogeneity [reviewed in 4]. A detailed understanding of how the neuroendocrine system affects the development and activity of different T-cell subsets is not yet available but there is evidence that hormones may play a more subtle role in regulating immunity than simply providing nonspecific immunosuppression [5, 6].

This paper describes studies on an experimental autoimmune disease, experimental allergic encephalomyelitis (EAE). There has been a number of previous reviews on this topic [7–12] but the aim of this present one is to describe what is known about the immunoregulatory mechanisms that control the development of EAE when this is elicited in the Lewis rat by the injection of a single purified autoantigen. While avoiding the potential difficulties of interpretation associated with having more than one encephalitogenic antigen, this system still poses fundamental questions about how the evolution of this disease is controlled by the experimental animal. In the course of these studies it became apparent that the neuroendocrine system played a crucial role in limiting the severity and duration of the disease and that genetic variation in this regulatory mechanism could determine whether a particular rat strain was susceptible to EAE. Furthermore, there was evidence that, although serum corticosterone levels were elevated only during the acute phase of the disease, such a transient could have a long-term effect on the ability of the encephalitogen-responsive T cells to express their disease-inducing potential.

Experimental Allergic Encephalomyelitis

Laboratory animals injected subcutaneously with a homogenate of spinal cord or brain in Freund's complete adjuvant (FCA) develop inflammatory infiltrates in the central nervous system (CNS) and varying degrees of paralysis. In its most severe form, the condition, known as EAE, is associated with areas of demyelination within the CNS and may prove fatal. EAE has been elicited in several species including monkeys, dogs, cats, rabbits, rats, mice and guinea-pigs but, despite this apparent lack of selectivity, studies with inbred strains of rats and mice have demonstrated a strong genetic component in susceptibility [13–16].

Experiments to determine which component of CNS tissue contains the encephalitogenic moiety have shown myelin basic protein (MBP) to be a major component in disease induction and that different portions of the molecule are encephalitogenic in different species [reviewed in 17]. MBP, an intracellular constituent of CNS and peripheral nerve myelin, has a molecular weight of 14–21.5 kD and is highly conserved between species. Variations in molecular weight arise through splice variation of mRNA but the significance of these variations is unknown [18, 19].

EAE in the Lewis Rat

In the susceptible Lewis strain of rat, the injection of purified guinea pig MBP in FCA gives rise to EAE in virtually all animals and, based on earlier work [20–23], it has been shown that a synthetic peptide of amino acid residues 70–86 of guinea pig MBP can elicit disease indistinguishable in time course and severity from that produced by the whole molecule [24, 25]. Signs of disease appear 10–12 days after immunization and develop in a character-istic sequence: first loss of muscle control of the tail followed by hind limb weakness that progresses to hind limbs paralysis in the majority of cases. Incontinence develops in a significant number of animals. The disease takes 2–3 days to reach its peak level of severity and then remits with those signs of disease that appear last being the first to go. By 6 days after disease onset recovery is virtually complete although mild relapses, about day 21 after immunization, occur in a minority of animals. Injection of MBP intraven-ously, or in FCA, about this time appears to increase the frequency of relapse [25]. However, by day 35 after immunization all rats are fully recovered and are completely resistant to attempts to induce further episodes of disease.

From the foregoing description of the evolution of EAE in the Lewis rat it appears that four phases can be distinguished, namely: the period between immunization and the onset of paralysis (phase 1); the period of acute paralysis (phase 2); the early post-recovery interval where relapse occasion-ally occurs (phase 3), and the final, indefinitely long period in which rats are refractory to the induction of further episodes of EAE (phase 4). The immunological processes underlying these four phases of disease are imper-fectly understood although substantial progress has been made. The follow-ing account attempts to review the major findings.

The Role of CD4+ and CD8+ Cells in the Pathogenesis of EAE

CD4+ T Cells

There are several sets of data that implicate CD4+ T cells in the pathogenesis of EAE. Splenocytes from rats with EAE, when cultured in vitro with Con A or MBP for 72 h and injected intravenously into naive syngeneic recipients, induce acute EAE (a condition known, rather inappropriately, as passive EAE) that develops in the hosts 4–5 days after cell transfer and lasts for a similar period of time [26, 27]. This transfer of EAE can be completely prevented if anti-CD4 monoclonal antibody (MAb) is included in the in vitro culture phase of splenocytes with MBP [28]. It had previously been shown that anti-CD4 MAb could virtually completely inhibit the mixed leukocyte response (MLR) in the rat [29] and the implication was that the in vitro phase of the transfer of passive EAE required the activation of CD4+ T cells.

Second, passive EAE was shown to be inducible in syngeneic Lewis strain recipients with T-cell lines and clones that were CD4+ [30, 31].

Third, active EAE (i.e. that induced by the injection of encephalitogen into the test animal) was shown to be completely preventable by the intravenous injection of anti-CD4 MAb [32] and this result was subsequently confirmed in studies on EAE in the mouse [33]. However, the experiments with the Lewis rat (unlike those in the mouse) used a MAb that did not deplete CD4+ T cells so the prevention of EAE in the rat could not be simply ascribed to ablation of CD4+ T cells. This point will be discussed again later in this review.

Although all these results showed that CD4+ T cells were essential for the development of EAE, they did not exclude the possibility that some other cell type was the actual effector cell that caused the nerve conduction defect. The mononuclear cell infiltrates in the CNS that are characteristic of EAE had led to the suggestion that the disease was a manifestation of a delayed-type hypersensitivity response to CNS antigens. However, it has been shown that severe passive EAE can be induced in recipient rats that have been given a lethal dose of ^{137}Cs γ-irradiation immediately before the injection of MBP-reactive splenocytes [34]. In these animals there was a 20-fold reduction in the number of cells infiltrating the CNS and the only cell type whose frequency in the lesions was unaffected by irradiation was that which expressed IL-2 receptors. It follows that the great majority of the cells that infiltrate the CNS in EAE are not essential for the development of paralysis and studies on the ability of encephalitogenic T-cell lines to destroy CNS vascular endothelial cell cultures pulsed with MBP strongly suggests that the CD4+ T cells themselves are the primary effectors of disease in this system [35]. Similar conclusions have been drawn for the effectors of skin allograft rejection when MHC incompatible grafts are involved [36]. T-cell lines that transfer passive EAE are potent producers of IFN-γ and TNF-α but whether these mediators are involved in pathogenesis is not clear. Cell lines that lack the ability to produce these mediators also lack encephalitogenic activity [37] but lines can lose the ability to induce EAE while still retaining the ability to produce IL-2, IFN-γ and TNF [38].

CD8+ T Cells

CD8+ T cells appear to play no essential role in EAE in the Lewis rat. As might be expected, CD8+ T cells can be found among the cells infiltrating the CNS but the depletion of CD8+ T cells by repeated injections of anti-CD8 MAb had no effect on the evolution of EAE in animals [39] and the observation, already alluded to, that severe EAE developed in irradiated rats injected with encephalitogenic CD4[+] T cells, despite the virtually complete absence of host T cells, was a strong indication that the CD8[+] T cells played no role in pathogenesis. However, the fact that active EAE, induced in rats depleted of CD8[+] T cells by MAb treatment, progressed through to the refractory phase implied that these cells were also not involved in establishing this phase of the disease [39]. Essentially the same result has been obtained in mice [40].

The Spontaneous Recovery Phase of EAE in Lewis Rats

As outlined in the Introduction, Lewis rats with EAE recover spontaneously within a few days of developing the first signs of paralysis and a number of hypotheses have been put forward to account for this recovery. One observation that proved invaluable in trying to understand this phase of the disease was that the onset of EAE is associated with a lymphopaenia that deepens as paralysis develops and remits as recovery occurs. This lymphopaenia is accompanied by a relative neutrophilia so that the total blood leukocyte count changes relatively little throughout the course of the disease. These changes in the leukocyte composition of the blood suggested that glucocorticoids might be involved and this possibility was amply verified by serum corticosterone determinations made during the course of the disease [41]. At the peak of disease, serum corticosterone levels in Lewis rats reach values of approximately 1 μM, i.e. at least 10 times basal levels at the nadir of the diurnal variation. Similar though rather smaller levels were recorded during passive EAE [41]. These high serum corticosterone levels were shown to be not simply an epiphenomenon since, in both active and passive EAE, adrenalectomy before the signs of paralysis resulted in complete abrogation of the spontaneous recovery phase of EAE and all animals developed progressive and fatal paralysis. Quantitative corticosterone replacement therapy, using steroid implants that mimicked the spontaneous release of corticosterone in intact rats, allowed adrenalectomized animals to recover from EAE with a tempo similar to that of normal animals. These results demonstrated that no other product of the adrenal glands played an essential role in the spontaneous recovery from EAE.

In reviewing the variety of explanations for the recovery of Lewis rats from EAE, it is evident that our own findings strongly supported earlier work by Levine et al. [42] who had concluded on the basis of their own studies that the spontaneous release of corticosterone from the adrenal glands of rats with EAE was crucial to their recovery. The work of Levine seems to have escaped the attention of most reviewers although it is a model of thoroughness and careful interpretation.

The Importance of Genetic Variation in the Stress Response

Lewis rats have been found to be the most useful strain in demonstrating a number of cell-mediated autoimmune diseases. Not only are they the strain of choice for EAE but also for experimental allergic neuritis, adjuvant and collagen-induced arthritis and experimental allergic uveitis. The biological reason for the unique position of the Lewis rat in these diseases became apparent when EAE was studied in another rat strain. PVG rats are consi-dered to be resistant to EAE though a low level of paralysis can sometimes be observed. Adrenalectomy of PVG rats renders them totally susceptible to EAE and all animals so treated develop severe progressive fatal paralysis just like Lewis strain rats [43]. Comparison of the corticosterone stress response in Lewis and PVG strain animals showed that the former produced lower levels of corticosterone and had smaller adrenal glands than sex- and age-matched PVGs. The importance of this genetically determined variation in the stress response has also been reported for the susceptibility to streptococ-cal cell wall arthritis where again Lewis rats are the most susceptible [44] and it has been suggested that similar genetic variation in man may be of immunological significance [45].

The Time Period in EAE for Which the Adrenal Glands Play Their Essential Regulatory Role

The ability of PVG rats to resist the induction of EAE unless adrenalec-tomized allowed a determination of the period over which the adrenals played their essential role. Adrenalectomy at the time of immunization with MBP/FCA or 8 days later resulted in fatal paralysis whereas delaying the operation to day 12 or later led to virtually no disease [43]. It seems from these results, and those in the Lewis rat, that the adrenals are required only until the initial elevation of corticosterone levels has occurred. The subse-quent refractory phase of EAE cannot, therefore, be ascribed to repeated elevations in corticosterone levels and it is evident that there is some long-term change in the immune status of rats that have recovered from EAE. However, this change in status is antigen-specific since rats that have recovered from EAE and have entered the refractory phase are still susceptible to experimental allergic neuritis (EAN) elicited by a peptide derived from the P2 protein of peripheral myelin. Similarly, refractoriness to EAN does not confer resistance to EAE [46].

The Refractory Phase of EAE

Suppression not Anergy

Animals that have recovered from EAE and that have become resistant to it do not lack cells that respond to MBP in vitro. These in vitro activated cells can passively transfer EAE to naive recipients indicating that the donor rats constitute an environment which prevents the realization of the encephalitogenic potential of these cells that they contain [47]. This conclusion also holds for T cells recovered from animals that have been treated with anti-CD4 MAb to prevent EAE [48]. Splenocytes from such MAb-treated rats, when activated in vitro with MBP, are particularly potent in their capacity to transfer EAE. Furthermore, using drugs to produce partial immunosuppression, it has been shown that EAE can be changed from a monophasic disease to a relapsing one [49, 50]. It appears that the immunosuppression interferes with the development of a mechanism that inhibits EAE and this result adds weight to the view that the refractory phase of EAE is an active process. In contrast, none of the data are consistent with the view that the refractory phase of EAE can be ascribed to anergy or deletion of cells with encephalitogenic potential. The situation appears similar to that of insulin-dependent diabetes where we have shown that rats possess diabetogenic cells but that these are inhibited in normal animals from causing β cell destruction [51]. Protection from diabetes is mediated by a TCR $\alpha\beta^+$ cell of CD45RC^{low} CD4$^+$ phenotype but the determination of the phenotype of the disease-preventing cell in EAE has been impeded by our inability to passively transfer the refractory status of EAE.

Nature of the Suppressive Mechanism

There is evidence that cell-mediated (CMI) and humoral responses (Ab) are reciprocally regulated. Immunization protocols that give vigorous CMI reactions may be associated with weak antibody responses and vice versa [52]. This phenomenon may be observed in EAE where the encephalitogenic peptide of guinea-pig MBP is covalently linked to bovine serum albumin. When emulsified in FCA, low doses (6 μg) of this conjugate induce severe EAE but little antipeptide antibody. At higher doses (60–100 μg) the severity of EAE is reduced significantly but the titre of antipeptide antibody increases [24]. This reciprocity of cell-mediated and humoral responses may be explained in terms of $CD4^+$ T-cell heterogeneity in that T cells that promote cell-mediated immunity produce lymphokines (in particular IFN-γ) that inhibit antibody responses while $CD4^+$ T cells that are most active in humoral immunity secrete lymphokines IL-4 and IL-10 which antagonize CMI [53]. It has been suggested that the refractory phase of EAE represents a change from a CMI response to the encephalitogen to a humoral one [45]. This conclusion is supported by several pieces of evidence: antibodies to the encephalitogenic peptide of MBP are detected at about the time that the refractory phase develops and in the rare animal that this antibody response occurs early the signs of EAE do not appear. Furthermore, as described, immunization protocols that result in high antipeptide antibody responses are associated with low levels of disease. Finally, there is the possibility that the elevated corticosterone levels associated with the disease influence the balance between CMI and humoral immunity as it has been shown that glucocorticoids promote the secretion of IL-4 while inhibiting that of IL-2 and IFN-γ [5, 54]. (It is notable in this regard that the $CD45RC^{low}$ $CD4^+$ cells that prevent diabetes in rats have the same phenotype as those known to produce IL-4.)

Prevention of EAE by Targeting the Encephalitogenic Peptide to B Cells

Treatment of mice with monoclonal anti-IgD results in polyclonal B-cell activation and elevated titres of serum IgG_1 [55]. These findings suggest that, as one may anticipate, presentation of antigen by B cells activates those T cells that are involved in providing B-cell help. Given the reciprocity that can occur between the strength of the CMI and humoral response to the same antigen [52] it may be further anticipated that coupling the encephalitogenic peptide of MBP to a monoclonal anti-IgD antibody should provide a reagent that can prevent EAE by favouring a humoral response to the peptide at the expense of the cell-mediated one.

This prediction has been amply verified: two intravenous injections of encephalitogenic peptide-anti-IgD conjugate given, respectively, on the day of immunization with MBP/FCA and 7 days later, can completely prevent EAE. Two different anti-IgD MAbs have been shown to be effective whereas free peptide, or peptide coupled to an irrelevant MAb of the same isotype, or free anti-IgD MAb were all nonprotective. Doses as low as 25 µg of peptide-anti-IgD conjugate have a detectable effect but for full protection doses of 100 µg are required. An exhaustive test of various protocols has not been made but a single 100-µg injection of conjugate, given 7 days before immunization with MBP/FCA, also seems to be protective [Day et al., submitted].

These results suggest a way in which other unwanted cell-mediated responses may be controlled while providing further indirect evidence that the refractory phase of EAE also reflects a qualitative change in the nature of the response to the encephalitogen. However, unless IL-4, and possibly IL-10, can be directly implicated, for example by abrogating the refractory phase by the use of MAbs to these lymphokines, the mechanism underlying the refractory phase of EAE will remain uncertain.

Acknowledgements

My thanks are due to my collaborators in the work described; to Ferenc Antoni, Steve Brostoff, Michael Day, Sheena Gowring, Nigel Groome, Iain MacPhee, Mike Puklavec, Jon Sedgewick, Steve Simmonds and Albert Tse and to Ros Sainty for typing the manuscript.

References

1 Munck A, Guyre PM, Holbrook NJ: Physiological function of glucocorticoids in stress and their relation to pharmacological actions. Endocr Rev 1984;5:25–44.
2 Sapolsky R, Rivier C, Yamamoto G, Plotsky P, Vale W: Interleukin-1 stimulates the secretion of hypothalamic corticotropin-releasing factor. Science 1987;238:522–524.
3 Berkenbosch F, van Oers J, del Rey A, Tilders F, Besedovsky H: Corticotropin-releasing factor-producing neurones in the rat activated by interleukin-1. Science 1987;238:524–526.
4 Coffman RL, Mosmann TR (eds): 35th Forum of Immunology. CD4+ T-Cell Subsets: Regulation of Differentiation and Function. Res Immunol 1991;142:1–84.
5 Daynes RA, Araneo BA: Contrasting effects of glucocorticoids on the capacity of T cells to produce the growth factors interleukin-2 and interleukin-4. Eur J Immunol 1989;19:2319–2325.
6 Daynes RA, Araneo BA, Dowell TA, Huang K, Dudley D: Regulation of murine lymphokine production in vivo. III. The lymphoid tissue microenvironment exerts regulatory influences over T-helper cell function. J Exp Med 1990;171:979–996.

7   Paterson PY: Experimental allergic encephalomyelitis and autoimmune disease. Adv Immunol 1966;5:131–208.

8   Weigle WO: Analysis of autoimmunity through experimental models of thyroiditis and allergic encephalomyelitis. Adv Immunol 1980;30:159–273.

9   Waksman BH: Current trends in multiple sclerosis research. Immunol Today 1981;2:87–93.

10 Raine CS: Biology of disease. Analysis of autoimmune demyelination: its impact upon multiple sclerosis. Lab Invest 1984;50:608–635.

11 Gonatas NK, Greene MI, Waksman BH: Genetic and molecular aspects of demyelination. Immunol Today 1986;7:121–126.

12 Zamvil SS, Steinman L: The T lymphocyte in experimental allergic encephalomyelitis. Annu Rev Immunol 1990;8:579–621.

13 Hughes RAC, Stedronska J: The susceptibility of rat strains to experimental allergic encephalomyelitis. Immunology 1973;24:879–884.

14 Williams RM, Moore MJ: Linkage of susceptibility of experimental allergic encephalomyelitis to the major histocompatibility locus in the rat. J Exp Med 1973;138:775–783.

15 Gasser DL, Hickey WF, Gonatas NK: The genes for nonsusceptibility to EAE in the Le-R and BH rat strains are not linked to RT1. Immunogenetics 1983;17:441–444.

16 Fritz RB, Skeen MJ, Jen Chou C-H, Garcia M, Egorov IK: Major histocompatibility complex-linked control of the murine response to myelin basic protein. J Immunol 1985;134:2328–2332.

17 Hashim GA: Myelin basic protein: structure, function and antigenic determinants. Immunol Rev 1978;39:60–107.

18 de Ferra A, Engh H, Hudson L, Kamholz J, Puckett C, Molineaux S, Lazzarini RA: Alternative splicing accounts for the four forms of myelin basic protein. Cell 1985;43:721–727.

19 Takahashi N, Roach A, Teplow DB, Prusiner SB, Hood L: Cloning and characterization of the myelin basic protein gene from mouse: one gene can encode both 14 Kd and 18.5 Kd MBPs by alternate use of exons. Cell 1985;42:139–148.

20 Martenson RE, Deibler GE, Kramer AJ, Levine S: Comparative studies of guinea pig and bovine myelin basic proteins. Partial characterization of chemically derived fragments and their encephalitogenic activities in Lewis rats. J Neuroimmunol 1975;24:173–182.

21 Kibler RF, Fritz RB, Chou C-HF, Jen Chou C-H, Peacocke NY, Brown NM, McFarlin DE: Immune response of Lewis rats to peptide C1 (residues 68–88) of guinea pig and rat myelin basic proteins. J Exp Med 1977;146:1323–1331.

22 Jen Chou C-H, Fritz RB, Chou C-HF, Kibler RF: The immune response of Lewis rats to peptide 68–88 of guinea pig myelin basic protein. 1. T cell determinants. J Immunol 1979;123:1540–1543.

23 Mannie MD, Paterson PY, U'Prichard DC, Flouret G: Induction of experimental allergic encephalomyelitis in Lewis rats with purified synthetic peptide: delineation of antigenic determinants for encephalitogenicity, in vitro activation of cell transfer and proliferation of lymphocytes. Proc Natl Acad Sci USA 1985;82:5515–5519.

24 MacPhee IA, Day MJ, Mason DW: The role of serum factors in the suppression of experimental allergic encephalomyelitis: evidence for immunoregulation by antibody to the encephalitogenic peptide. Immunology 1990;70:527–534.

25 MacPhee IAM, Mason DW: Studies on the refractoriness to reinduction of experimental allergic encephalomyelitis in Lewis rats that have recovered from one episode of the disease. J Neuroimmunol 1990;27:9–19.

26 Panitch HS, McFarlin DE: Experimental allergic encephalomyelitis: enhancement of cell-mediated transfer by concanavalin A. J Immunol 1977;119:1134–1137.

27 Richert JR, Driscoll BF, Kies MW, Alvord ECJ: Adoptive transfer of experimental allergic encephalomyelitis: incubation of rat spleen cells with specific antigen. J Immunol 1979;122:494–496.

28 Swanborg RH: Autoimmune effector cells. V. A monoclonal antibody specific for rat helper T lymphocytes inhibits adoptive transfer of autoimmune encephalomyelitis. J Immunol 1983;130:1503–1505.

29 Webb M, Mason DW, Williams AF: Inhibition of mixed lymphocyte response by monoclonal antibody specific for a rat T lymphocyte subset. Nature 1979;282:841–843.

30 Ben-Nun A, Cohen IR: Experimental autoimmune enccphalomyelitis mediated by T cell lines: process of selection of lines and characterization of the cells. J Immunol 1982;129:303–308.

31 Sedgwick JD, MacPhee IAM, Puklavec M: Isolation of encephalitogenic CD4+ T cell clones in the rat. Cloning methodology and interferon-γ secretion. J Immunol Methods 1989;121:185–186.

32 Brostoff SW, Mason DW: Experimental allergic encephalomyelitis: successful treatment in vivo with a monoclonal antibody that recognizes T helper cells. J Immunol 1984;133:1938–1942.

33 Waldor MK, Sriram S, Hardy R, Herzenberg LA, Lanier L, Lim M, Steinman L: Reversal of experimental allergic encephalomyelitis with a monoclonal antibody to a T-cell subset marker. Science 1985;227:415–417.

34 Sedgwick J, Brostoff S, Mason D: Experimental allergic encephalomyelitis in the absence of a classical delayed type hypersensitivity reaction. Severe paralytic disease correlates with the presence of interleukin-2 receptor-positive cells infiltrating the central nervous system. J Exp Med 1987;165:1058–1075.

35 Sedgwick JD, Hughes CC, Male DK, MacPhee IA, ter Meulen V: Antigen-specific damage to brain vascular endothelial cells mediated by encephalitogenic and non-encephalitogenic CD4+ T cell lines in vitro. J Immunol 1990;145:2474–2481.

36 Dallman MJ, Mason DW: Cellular mechanisms of skin allograft rejection in the rat. Transplant Proc 1983;15:335–338.

37 Powell MB, Mitchell D, Lederman J, Buckmeier J, Zamvil SS, Graham M, Ruddle NH, Steinman L: Lymphotoxin and tumor necrosis factor-α production by myelin basic protein-specific T cell clones correlates with encephalitogenicity. Int Immunol 1990;2:539–544.

38 Day MJ, Mason DW: Loss of encephalogenicity of a myelin basic protein-specific T cell line is associated with a phenotypic change but not with alteration in production of interleukin-2, γ-interferon or tumor necrosis factor. J Neuroimmunol 1990;30:53–59.

39 Sedgwick JD: Long-term depletion of CD8+ T cells in vivo in the rat: no observed role for CD8+ (cytotoxic/suppressor) cells in the immunoregulation of experimental allergic encephalomyelitis. Eur J Immunol 1988;18:495–502.

40 Sriram S, Carroll L: In vivo depletion of Lyt-2 cells fails to alter acute and relapsing EAE. J Neuroimmunol 1988;17:147–157.

41   MacPhee IA, Antoni FA, Mason DW: Spontaneous recovery of rats from experimen-
     tal allergic encephalomyelitis is dependent on regulation of the immune system by
     endogenous adrenal corticosteroids. J Exp Med 1989;169:431–445.
42   Levine S, Sowinski R, Steinetz B: Effects of experimental allergic encephalomyelitis
     on thymus and adrenal: relation to remission and relapse. Proc Soc Exp Biol Med
     1980;165:218–224.
43   Mason D, MacPhee I, Antoni F: The role of the neuroendocrine system in determin-
     ing genetic susceptibility to experimental allergic encephalomyelitis in the rat.
     Immunology 1990;70:1–5.
44   Sternberg EM, Hill JM, Chrousos GP, Kamilaris T, Listwak SJ, Gold PW, Wilder
     RL: Inflammatory mediator-induced hypothalamic-pituitary-adrenal axis activation
     is defective in streptococcal cell wall arthritis-susceptible Lewis rats. Proc Natl Acad
     Sci USA 1989;86:2374–2378.
45   Mason DW: Genetic variation in the stress response: susceptibility to experimental
     allergic encephalomyelitis and implications for human inflammatory disease. Imm-
     unol Today 1991;12:57–60.
46   Day MJ, Tse A, Mason DW: The refractory phase of experimental allergic encepha-
     lomyelitis in the Lewis rat is antigen-specific in its induction but not in its effect.
     J Neuroimmunol 1991;34:197–203.
47   Holda JH, Welch AM, Swanborg RH: Autoimmune effector cells. 1. Transfer of
     experimental encephalomyelitis with lymphoid cells cultured with antigen. Eur J
     Immunol 1980;10:657–659.
48   Sedgwick JD, Mason DW: The mechanism of inhibition of experimental allergic
     encephalomyelitis in the rat by monoclonal antibody against CD4. J Neuroimmunol
     1986;13:217–232.
49   Polman CH, Matthaei I, de Groot CJA, Koetsier JC, Sminia T, Dijkstra CD: Low
     dose cyclosporin A induces relapsing remitting experimental allergic encephalo-
     myelitis in the Lewis rat. J Neuroimmunol 1988;17:209–216.
50   Minagawa H, Takenaka A, Itoyama Y, Mori R: Experimental allergic encephalomy-
     elitis in the Lewis rat. A model of predictable relapse by cyclophosphamide. J Neurol
     Sci 1987;78:225–235.
51   Fowell D, McKnight AJ, Powrie F, Dyke R, Mason D: Subsets of CD4+ T cells and their
     roles in the induction and prevention of autoimmunity. Immunol Rev 1991;123:37–64.
52   Parish CR, Liew FY: Immune response to chemically modified flagellin. 3. Enhanced
     cell-mediated immunity during high and low zone antibody tolerance to flagellin.
     J Exp Med 1972;135:298–311.
53   Mosmann TR: Cytokine secretion phenotypes of TH cells: How many subsets, how
     much regulation? Res Immunol 1991;142:9–13.
54   Daynes RA, Meikle AW, Araneo BA: Locally active steroid hormones may facilitate
     compartmentalization of immunity by regulating the types of lymphokines produced
     by helper T cells. Res Immunol 1991;142:40–45.
55   Goroff DK, Holmes JM, Bazin H, Nisol F, Finkelman FD: Polyclonal activation of
     the murine system by an antibody to IgD. IX. Contribution of membrane IgD cross-
     linking to the generation of an in vivo polyclonal antibody response. J Immunol
     1991;146:18–25.

Don Mason, MD, MRC Cellular Immunology Unit,
Sir William Dunn School of Pathology, Oxford OX1 3QU (UK)

Coffman RL (ed): Regulation and Functional Significance of T-Cell Subsets.
Chem Immunol. Basel, Karger, 1992, vol 54, pp 72–102

Postthymic Differentiation of CD4 T Lymphocytes: Naive Versus Memory Subsets and Further Specialization among Memory Cells

Kevin J. Horgan, Yoshiya Tanaka, Stephen Shaw

Experimental Immunology Branch, National Cancer Institute,
National Institutes of Health, Bethesda, Md., USA

Introduction

Much of the excitement regarding T-cell differentiation has focused on events in the thymus. However, we view postthymic T-cell differentiation to be as complex and at least as important. The most obvious reflection of the complexity of postthymic T-cell differentiation is the startling heterogeneity in surface phenotype of peripheral T cells. Within T cells the distinction between CD4 and CD8 is the most familiar and best understood dichotomy [1]. However, within either CD4 or CD8 subsets there is also an enormous complexity/heterogeneity in expression of a multitude of other surface markers. We began grappling with these issues five years ago in an attempt to understand the significance of our unexpected finding that LFA-3 (CD58) is differentially expressed on subsets of both CD4 and CD8 cells [2]. Our analysis commenced with CD4 cells for two reasons: (1) in view of their abundance it would be easier to study them; (2) we had the impression, which in retrospect seems absurd, that CD4 cells (unlike CD8 cells and their melange of NK relatives) were relatively homogeneous and could be quickly dealt with. Five years later we have a huge amount of data, largely unpublished, which indicates very complex heterogeneity among both CD4 and CD8 cells. The challenge now is to find simplifying/unifying principles which may clarify the process of peripheral T-cell differentiation and make the morass of complex data comprehensible.

The most important simplifying principle that has emerged from our and others' studies is the concept of naive and memory T cells. Reduced to the bare essentials, the concepts are that: (1) reciprocal expression of different isoforms of the CD45 molecule identifies two subsets of peripheral T cells generally known as naive and memory cells; (2) naive and memory cells have fundamental differences in surface phenotype, activation requirement, effector function, and in vivo migration. The bulk of this review deals with these issues for CD4⁺ T cells.

We propose a second simplifying principle, that naive CD4⁺ T cells are relatively homogeneous phenotypically, while memory CD4⁺ T cells have a great deal of additional heterogeneity. We interpret our data to mean that there is a large pool of rather homogeneous naive CD4⁺ T cells which are awaiting their first stimulation in the periphery. Following activation they undergo complex differentiation events with accompanying changes in phenotype. Many of the phenotypic changes will be present in all memory cells. However, some of the phenotypic changes will be idiosyncratic to particular circumstances of T-cell activation; the resulting memory T cell will possess a molecular memory of relevant details of its activation such as the organ in which it was stimulated. Each of many kinds of memory cells will be endowed with specialized capacities in homing, cell collaboration and effector function. The latter portion of this review briefly surveys memory cell heterogeneity (concentrating on our studies of CD45RB and integrin markers VLA-α4/VLA-β1) and expands on this proposed model of memory cells as 'an army of specialists'.

Naive and Memory Cells

As mentioned above, subsets of T cells can be identified that appear to represent distinct stages in the postthymic differentiation of T cells [3–5]. Operationally, the most reliable markers which distinguish these subsets are different forms of the CD45 molecule (formerly known as T200 or the leukocyte common antigen). This, coupled with the fact that CD45 is a protein tyrosine phosphatase whose function is critical to peripheral T-cell activation (see below), assures CD45 a prominent place in discussions of T-cell subsets; excellent reviews are available to supplement the following rudimentary description. CD45 is a family of glycoproteins present on all nucleated hemopoietic cells which constitute approximately 10% of the surface protein of the cell [6]. The isoforms that are differentially expressed

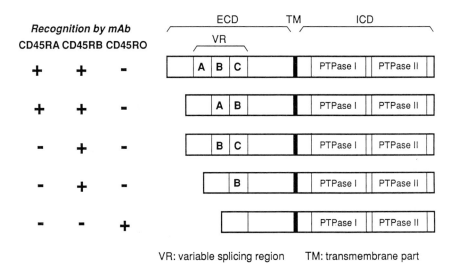

Fig. 1. Schematic of CD45 molecule.

on subsets of T cells are distinguished by variability in the structure of the extracellular part of the molecule. This variability results in part from differential RNA splicing of transcripts of a single gene; the isoforms differ with respect to the inclusion or exclusion of 3 exons (A, B, and C) which code for amino acids near the amino-terminal extracellular end (fig. 1) [7–10].

Three kinds of human CD45 isoform-specific mAbs have been studied most extensively. Isoforms which include the A exon (220 and 205 kD) are recognized by antibodies known as CD45RA. Isoforms which include the B exon (220, 205 and 190 kD) are reactive with CD45RB antibodies. The lowest molecular weight isoform (180 kD) which excludes the A, B and C exons is recognized by CD45RO antibodies [11, 12].

Of fundamental importance is the reciprocal expression of the CD45RA and CD45RO determinants in resting populations of CD4+ T cells which allows the definition of two subsets (fig. 2) [13, 14]. The prevailing concept is that T cells expressing CD45RA are naive in the sense that they have not been activated following export from the thymus. The reciprocal subset (CD45RA-/CD45RO+) are cells which have been previously activated, generally by encounter with specific antigen, and have reverted back to a resting

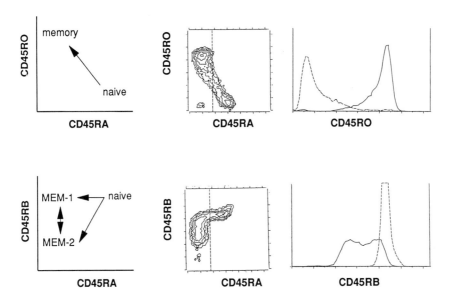

Fig. 2. Identification of subsets with CD45 isoforms. The top row of figures illustrates the relationship between CD45RA and CD45RO. The contour plot in the middle shows a clear reciprocal relationship between CD45RA and CD45RO in resting CD4+ T cells. The idealized schema to the left gives labels to the subsets shown. The histograms on the right show more precisely the expression of CD45RO on CD45RA+ cells (dotted line) and CD45RA⁻ cells (solid line). The bottom row of figures illustrates the relationship between CD45RB and CD45RA. The contour plot in the middle shows three populations of cells identified. The idealized schema to the left gives labels to the subsets shown. The histograms on the right show the homogeneous high expression of CD45RB on CD45RA+ naive cells and the subsets identified by CD45RB among CD45RA⁻ memory cells.

state by many criteria; these are designated memory cells. Thus, for the purpose of our discussion, naive cells are CD45RA+/CD45RO⁻ and memory cells are CD45RA⁻/CD45RO+ [2, 13, 15, 16]. Though other molecules have been employed as markers of naive and memory cells (see below) the clean reciprocity of CD45RA and CD45RO expression is unique and for that reason we employ them as the benchmark determinants for provisionally identifying naive and memory cells.

The evidence supporting the conclusion that these are in fact naive and memory cells is derived from several observations:

(1) The vast majority of neonatal cord T cells are of naive phenotype (they express CD45RA) and with aging there is a progressive increase of cells

of memory (CD45RO$^+$) phenotype present in the circulation. This is consistent with the belief that a neonate has had minimal antigenic exposure and that there is a gradual increase in the proportion of circulating memory cells as a consequence of continued exposure to novel antigens over the lifetime of the individual [2, 17, 18].

(2) Memory cells differentiate from naive cells in vivo following antigen exposure. The in vitro corollary is that following activation with a variety of mitogens, separated naive cells acquire a memory phenotype; the reverse conversion has not been seen [2, 16]. Of note, this conversion has been documented not only in bulk cultures but also by limiting dilution analysis [5, 15].

(3) A 'recall' response is the immunologic response of an individual to a specific antigen to which previous exposure has occurred. Typically, T-cell proliferation to antigens like tetanus toxoid, *Candida albicans* or influenza occurs only with cells from individuals previously exposed to these antigens. A putative marker of either naive or memory cells should allow separation of T cells into one subset with a strong recall antigen response and another without a response. The criteria is fulfilled by the CD45RA and CD45RO isoforms [2, 15].

Complexity Beneath the Surface

The foregoing paradigm of naive and memory cells has proved to be a very useful one, but it is undoubtedly an oversimplification. Before developing the concept further, we alert the reader to a number of observations that do not fit neatly into the naive/memory paradigm and illustrate gaps in our understanding.

(1) Various lines of evidence suggest that neither 'memory' nor 'naive' cells are functionally homogeneous. The issue of memory cell heterogeneity is dealt with in the second half of the review. In addition, there is functional heterogeneity among the CD45RA$^+$ 'naive' cells. Kulova et al. [19] have demonstrated that a subset of CD45RA$^+$ cells are unresponsive to activation via CD3 cross-linking. They propose that this may represent a functionally immature T-cell pool which is self-regenerating but able to differentiate into functionally mature naive cells.

(2) Byrne et al. [20] found numerous CD45RO$^+$ T cells in human fetal spleen and there are many CD45RO$^+$ cells in human thymus; it has been suggested that the expression of CD45RO by thymocytes identifies those cells

destined for intrathymic death [21–23]. It is likely that the expression of the CD45RO isoform in these circumstances has a different physiologic significance than the expression of CD45RO by circulating T cells.

(3) We have found that in some samples of cord blood a significant proportion of the cells have dull rather than bright expression of CD45RA, which is not typical for naive cells. Perhaps these cells are a reflection of in utero antigen exposure or alternatively antigen exposure may not be necessary in some circumstances for effecting a phenotypic conversion from CD45RA+ to CD45RA− among peripheral T cells.

(4) A provocative finding has been the demonstration that in some in vitro culture conditions, activated CD45RA+ cells continue to express CD45RA [24]. This violates the dogma that loss of CD45RA always accompanies peripheral T-cell activation. Although it remains to be seen how this relates to phenotypes in vivo, it illustrates that alternative splicing of CD45 is regulated in a complex manner.

(5) Another complexity is the existence of donor-dependent variation in the identification of CD45 isoforms [25, 26]. Using a variety of CD45RA mAbs one study reported 8% of healthy donors had no CD45RA− cells and this phenotypic profile was inherited in an autosomal-dominant pattern [26]. Interestingly the profile for CD45RO expression was not unusual. In our unpublished studies we have occasionally seen donors with an apparent lack of CD45RA− cells but when we stain the cells with an alternative CD45RA antibody a CD45RA− population is clearly identified. These findings raise the important issues that not all CD45RA mAbs are equivalent and that CD45 isoforms may differ not only in protein sequence but also in glycosylation. Some CD45RA antibodies detect protein determinants while others detect carbohydrate determinants [27, 28]. Since dramatic differences in reactivity with CD45RA mAbs can be explained by variations in glycosylation, the apparent dearth of CD45RA− cells in some donors with certain CD45RA mAbs may simply represent an idiosyncrasy of glycosylation of the CD45RA isoform.

(6) Complexities are amplified by comparisons between species and between isoforms. There are many parallels between findings with high molecular weight CD45 isoforms in the human (CD45RA) and high molecular weight isoforms in the rat detected by OX-22, which proves to be CD45RC-specific [M. McCall and A.W. Barclay, submitted]. As a result it is tempting to generalize from one to the other; however, this is unwarranted because of the difference between isoforms and potentially between species. A recent study demonstrates that OX-22− cells revert to an OX-22+ pheno-

type in adoptive transfer experiments in rats [29, 30]; in contrast to the interpretation given by the authors, we view these data as not relevant to the model that CD45RA and CD45RO distinguish naive and memory cells in humans.

Having detailed some complexities, we will now review some of the large amount of robust data delineating the features of naive and memory cells as defined by our original criteria of reciprocal expression of CD45RA and CD45RO.

Naive and Memory Cells; Phenotype and Function

Naive and memory cells differ not only in their expression of CD45RA and CD45RO, but also in their expression of a number of other molecules. Inspection of data summarized in table 1 shows that a number of molecules are increased on memory cells, and that many of them are known to be functionally important as cell adhesion molecules [2, 31–34]. These molecules include CD2, LFA-3 (CD58), ICAM-1 (CD54), CD29 (also known as 4B4 the VLA-β1 chain), VLA-4 (CD49d), VLA-5 (CD49e) and VLA-6 (CD49f) and CD26 (Ta1). Although CD44 is also expressed more by memory cells than naive [2], our subsequent surveys of more molecules suggests that the small increase of CD44 is relatively nonspecific and shared by many other molecules. Based on their differential expression of this array of adhesion molecules, naive and memory cells would be expected to differ substantially in their adhesion capacity; this prediction is confirmed by a variety of in vivo and in vitro observations (see below).

Within this category of molecules preferentially expressed on memory cells, there is considerable diversity with respect to level of expression on naive cells and heterogeneity among memory cells. For most of these molecules there is simply greater expression on memory compared to naive cells [2]; however, for others, such as LFA-3 and ICAM-1, there is usually no detectable expression on naive cells but modest expression on memory cells [2]. Most of the molecules listed are expressed at relatively homogeneous levels on memory cells; however, more careful analysis reveals that many of them show heterogeneity in level of expression among memory cells. This is particularly true of the VLA integrins (particularly VLA-4), HECA-452 and CD26; this issue is dealt with in the second half of the review.

It is noteworthy that several molecules regarded as being markers of T-cell activation have modest expression on memory cells, suggesting that

Table 1. Molecules differentially expressed on resting CD4+ naive and memory cells

Molecule	Other names	Approximate ratio memory:naive	Comments
Integrin family			
LFA-1	CD11a/CD18	2:1	Receptor for ICAM-1 and ICAM-2 mediates leukocyte adhesion and signaling CD18 is common β2 chain
VLA-β	CD29	4:1	Common β1 chain of VLA subfamily integrin ? signaling
VLA-3α	CD49c	6:1	Ligands collagen, FN and LN
VLA-4α	CDw49d	3:1[1]	Ligands FN and VCAM-1 roles in T-cell homing and ? signaling
VLA-5α	FNR CDw49e	3:1	Ligand FN, ? signaling
VLA-6α	CDw49f	3:1	Ligand LN, ? signaling
Immunoglobulin supergene family			
ICAM-1	CD54	3:1	Ligand for LFA-1 leukocyte adhesion and signaling
CD2	T11	2:1	Receptor for LFA-3 T-cell adhesion and signaling
LFA-3	CD58	10:1	Ligand for CD2 T-cell adhesion and signaling
Other adhesion or homing-related receptors			
CD44	Hermes Pgp-1 ECMR III	1.5:1	T-cell homing, adhesion to erythrocytes and signaling
CD26	DPP IV	2:1[1]	? ligand FN or collagen
HECA-452	CLA	4:1[1]	SLe[a], SLe[x], ligand ELAM-1 ? skin-associated homing
Control			
CD3	TCR	0.9:1	Antigen-specific T-cell receptor complex
CD45	LCA, T200	1.2:1	Invariant part of CD45 molecule cytoplasmic portion tyrosine phosphatase

[1] Clear heterogeneity within CD45RO+ memory cells.

the cells may be in a state of low grade activation. These include both the α and β chains of the IL-2 receptor and ICAM-1 [35–37]. Our own unpublished data also shows modest expression of these molecules on memory cells. In addition, memory cells are also slightly larger than naive cells which also suggests that they may be in a somewhat activated state [35]. This is supported by data from the sheep that memory cells are cycling more rapidly than naive cells [38]. Beverley [39] has suggested that this low grade activation state may be a consequence of continued antigenic stimulation which may be necessary for the maintenance of immunologic memory.

There are marked differences in the in vivo activation requirements of separated naive and memory cells. One caveat to note before summarizing the results is that some results are very dependent on the details of the experimental protocol used. In particular, it is vital where possible to use negative selection to isolate the cells in order to minimize perturbing the functional properties of the cells. Several studies have demonstrated that T cells separated by positive selection methods such as rosetting have altered functional properties because they are partially activated [40–43].

The proliferative response of naive cells is generally less vigorous than that of memory cells in the commonly used in vitro models of T-cell activation using CD2 mAb pairs, CD3 mAbs and staphylococcal enterotoxin A (SEA) [44–47]. The magnitude of this difference is most pronounced at low doses of the activating stimulus [45]. Such responses are considered to be of importance not only because they model the most fundamental property of the T cell which is its capacity to be activated by engagement of the T-cell receptor but also may explain in part why a primary immune response is less vigorous than a secondary immune response. The response of naive cells is also suboptimal when a monocyte-independent system is used [45]. This can be done with certain CD3 mAbs such as 64.1 that can provide a proliferative signal when immobilized on plastic in the absence of antigen-presenting cells (APC). Interestingly, the discrepancy of the naive and memory response is not so marked as in a monocyte-dependent system. This implies that a component of the differential response of naive and memory cells in monocyte-dependent systems is a suboptimal interaction of the naive cell with the APC, in part because of lower expression of adhesion molecules [45].

It is important to emphasize that naive cells can proliferate as well as (if not better than) memory cells when the appropriate signal(s) is provided. This is true for high doses of CD3 mAb, SEA and most consistently high dose PHA. At low doses of CD3 and SEA the naive cell response can be preferentially augmented by the provision of appropriate costimuli such as mAbs against CD28, CD44 and CD45 [45, 48]. The naive cell response to CD2 mAb pairs is particularly well augmented by mAbs against CD45. The physiologic relevance of these data may be that the provision of appropriate costimuli is critical in determining whether a naive cell will be activated when a novel antigen is encountered. We believe that the secondary lymphoid organs, notably lymph node, provide an optimal environment for such naive cell stimulation in vivo (see below).

What Is the Biochemical Basis of the Differential Responsiveness?

The cytoplasmic portion of CD45 is a protein tyrosine phosphatase [49]; it is likely that this is relevant to the differential responsiveness of naive and memory cells, as many cellular events are regulated by tyrosine phosphorylation/dephosphorylation events. CD45 is essential for antigen-induced proliferation as well as for activation via CD2 [50, 51]. The tyrosine kinase p56^{lck} which is physically associated with CD4 and CD8 and believed to be a pivotal element in the molecular events of antigen-induced T-cell activation [52] may be a substrate for the CD45 tyrosine phosphatase [53, 54]. This raises the possibility that differential engage-ment/localization of the CD45 tyrosine phosphatase in naive and memo-ry cells may be an important determinant of the difference in activation thresholds of the subsets by altered interactions with p56lck in naive and memory cells respectively [55]. It is widely speculated that the activity of CD45 in naive and memory cells is influenced by differential reactivity of their CD45 isoforms with ligands. Theoretically, ligands may exist on the apposing cell ('trans') or on molecules on the same cell ('cis'); the first ligand to be identified is CD22 [56]. Studies using CD45 antibodies in vitro to perturb T-cell activation have shown quite complex results de-pending on the details of the system employed [57–60]; this is fully consistent with the complex regulatory roles which tyrosine phosphoryla-tion plays in activation events. One particularly interesting finding is the marked costimulatory effect of CD45 mAbs on CD2-mediated activation of naive cells especially in light of the subsequent demonstration that CD2 and CD45 are physically associated on the T-cell surface [58].

Another critical difference between naive and memory cells is the profile of cytokines that they secrete. Numerous studies have produced the consistent finding that naive cells produce IL-2, and small amounts of IFN-γ but no IL-4, IL-5 or IL-6, whereas memory cells produce signifi-cant amounts of all five cytokines [2, 61–65]. Some groups have reported greater production of IL-2 by naive cells but this has not been a con-sistent finding [2, 14, 61, 62, 66]; this apparent conflict may result from different kinetics of production in the various model systems [61]. It is important to emphasize that the naive cell cytokine profile is restricted to production of only IL-2 and minimal IFN-γ even when the cells have been stimulated to proliferate more than memory cells. Thus, an important part of the efficiency of an anamnestic response may be that a memory cell is capable of promptly secreting the necessary effector lymphokines to coordinate an optimal immune response. It is noteworthy that the cytokine profile of human naive or memory cells corresponds to neither murine TH1 nor TH2 subsets which are defined by their cytokine profile; TH1 cells produce IL-2 and IFN and TH2 cells produce IL-4 and IL-5 [67–70]. We emphasize that TH1 and TH2 cells both represent memory cell subsets. There has been a single report of differential expression of CD45 isoforms in the murine TH1 and TH2 clones [71] but the significance of this is unclear. Very recently several reports of human T cell clones with the cytokine profiles of murine TH1 and TH2 clones have appeared, though without phenotypic characterization [69, 70, 72, 73].

As well as having a different profile of cytokine production, naive and memory cells differ in their response to various cytokines [74–78]. IL-4 has an inhibitory effect on memory cell proliferation whereas naive cell proliferation is enhanced [77]. Memory cells are preferentially costimulated by IL-1 [79]. Both subsets' proliferative response to CD2 is enhanced by IL-6 [74, 76]. A potential explanation for the poorer responsiveness of naive cells to CD2 may be their inability to produce IL-6; when exogenous IL-6 is added, naive cells proliferate in response to CD2 [76]. A recently described cytokine of the RANTES family causes the selective migration of only memory cells [78]. This would be expected to complement the selective migration enabled by differential expression of adhesion molecules and facilitate the rapid entry of memory cells into tissue and areas of inflammation (see below).

The original description of these subsets was done using CD45RA (2H4 mAb) and CD29 (4B4). In those studies CD45RA$^+$CD29$^-$ cells were designated suppressor-inducer cells because they induced CD8$^+$ T cells to inhibit pokeweed mitogen (PWM)-derived B-cell differentiation [80] and CD45RA$^-$CD29$^+$ were designated as helper-inducer cells because they provided help for B-cell differentiation induced by PWM [81]. Although the functional properties of these subsets have been confirmed by others it is clearly not as simple as originally presented. Two studies showed that both naive and memory cells could help or suppress B-cell responses, with the nature of the activating stimulus determining the nature of the response [82, 83].

In the context of the naive/memory cell hypothesis, it is reassuring that the properties of naive cells when isolated from healthy adults are generally very similar to those of T cells obtained from neonatal cord blood. This has been studied in several different ways in terms of response not only to activation via CD2 and via CD3 but also in production of cytokines [2, 84–86]. The studies of Ehlers and Smith [85] are worthy of comment even though they did not report phenotypic characterization of the cell populations they used. They found that cord blood cells were capable of producing mRNA for only IFN-γ and IL-2. However, with secondary stimulation the same cells produced IL-4 and IL-5 mRNA, implying that they had now converted into memory cells.

Tissue Localization

One of the most satisfying elements in the understanding of naive and memory cells is the convergence of information regarding their adhesive function and their tissue localization. In accordance with their enhanced expression of defined adhesion molecules, memory cells adhere better than naive cells to relevant purified ligands immobilized on plastic such as VCAM-1 (via VLA-4), ICAM-1 (via LFA-1), fibronectin (via VLA-5 and VLA-4) and laminin (via VLA-6) [33, 87, 88]. Furthermore, preferential memory cell adhesion is also seen to endothelial cells via VLA-4 binding to VCAM-1 and LFA-1 binding to ICAM-1 and ICAM-2 [88, 89]. The relevance of these interactions is that they begin to explain the selective migration of memory cells out of the circulation into different tissues and anatomic compartments.

The enhanced expression of adhesion molecules that mediate binding to endothelium (VLA-4, LFA-1 and ELAM-1 ligand) and to extracellular matrix (VLA-4, VLA-5 and VLA-6) by memory cells likely provides the molecular basis for the preferential migration of memory cells into tissue and into sites of inflammation [33, 87, 90]. In skin, bronchial mucosa, and gut the T cells are of memory phenotype [91–94] according to their expression of CD45 isoforms. The studies on an induced skin blister are of particular interest as they illustrate the prompt entry of memory cells rather than in situ conversion of naive cells into memory cells [95]. Breast milk T cells are also of memory phenotype which may be of particular importance in the transfer of maternal immunity to the infant [96].

What Are the Recirculation Patterns and Sites of Activation of
Naive Cells?

Elegant studies in the sheep help address that question. In sheep, as in humans, CD45 isoforms subdivide naive and memory cells, and memory cells preferentially express adhesion molecules such as LFA-1 and CD2 [38]. Two routes of lymphocyte recirculation through the lymph node were distinguished in sheep: (1) naive cells enter the lymph node directly from peripheral blood via high endothelial venules (HEV) and leave the lymph node via efferent lymphatics to re-enter the blood; and (2) memory cells selectively traffic from blood to tissue and drain to the local lymph nodes via afferent lymphatics [38]. These findings corroborate the human data discussed in the preceding paragraph that the T cells in tissue are over-whelmingly of memory phenotype [91–93].

The sheep studies highlight the critical role of lymph nodes in recirculation of naive cells. The lymph node serves as a specialized site that brings together the rare relevant naive T cells, specialized APC, and the antigen load drained from local tissue in a microenvironment particularly suited to T-cell stimulation. Our understanding is that naive cells are typically sensitized, not at the site of antigen entry where there are very few present, but rather in the regional lymph nodes [97–99]. For example, during skin sensitization, the local specialized (APCs) take up antigen and then migrate to the regional lymph nodes. There antigen is presented to the naive T cells which are passing through in great numbers. It is essential that there is a large migration of naive cells and a high frequency of encounter between them and APC to ensure that the rare naive cell specific for that antigen will be triggered. Of additional importance, the lymph node milieu may provide a multitude of accessory signals derived from cytokines, extracellular matrix and/or cell surface ligands to facilitate naive cell activation [97]. In support of this scenario of events, many cells are seen in lymph node, tonsil and in the lymphoid aggregates of gut-associated lymphoid tissue that are simultaneously CD45RA+ and CD45RO+ suggesting that they are in a transitional state from naive to memory following activation by specific antigen [100]. Furthermore, predictable patterns of emergence of IL-2 producing T cells can be followed in the lymph node during antigen sensitization [99].

Given the selective migration of naive cells into lymph node it is to be expected that adhesion molecules relevant to that process are enriched on naive cells (in distinction to the adhesion molecules discussed above). Indeed naive cells preferentially express the homing receptor LECAM-1 (LAM-1,

Leu-8, Mel-14, Ly-22) which is believed to mediate adhesion to peripheral lymph node HEV and thereby allow migration from blood by HEV into lymph node [101, 102]. The few other known molecules that are preferentially expressed on naive cells include CD7 and CD31, which appear to mediate selective naive cell adhesion and/or activation [103–105].

Disease States

Although phenotyping of patient material for CD45RA and CD45RO isoforms has become increasingly common, for the most part we do not understand the fundamental biology which underlies the alterations observed. The one notable exception is that memory cells predominate in areas of chronic inflammation; this follows from what we know about the adhesion molecules which are induced on endothelium during inflammation (VCAM-1, ICAM-1, ELAM-1) and about memory cell predominance of the T-cell ligands with which they interact (VLA-4, LFA-1 and specific carbohydrate) [106]. Examples of this include the lesions of tuberculoid leprosy where many memory cells and strong antigen reactivity exist in contrast to the ineffective immune response of lepromatous leprosy where the lesions are devoid of memory cells [107]. Memory cells are prevalent also in tuberculous pleuritis, chronic synovitis and in inflammatory cerebrospinal fluid [108–112]. Tumor-infiltrating lymphocytes are mostly memory cells [113]. This preponderance may reflect not only influx of memory cells but also ongoing conversion of naive cells into memory cells.

Increases in the proportion of circulating T cells of memory phenotype has been seen in several diseases of apparent autoimmune etiology: systemic lupus erythematosus (SLE), Sjögren's syndrome, rheumatoid arthritis (RA), thyrotoxicosis, idiopathic thrombocytopenic purpura (ITP) and multiple sclerosis (MS) [114–119]. Two explanations have been suggested to account for these observations. One is that the 'imbalance' is of etiologic importance and that the decrease of CD45RA+ cells which have suppressor function allows excessive responses to the offending antigen [115]. The other hypothesis is that the alteration may be secondary to the ongoing immunologic reactions which result from the disease state [118]. It is likely than both explanations are relevant in part. There is some evidence to suggest that fluctuations in the proportion of circulating naive and memory cells may correlate with relapse in MS [120] and disease progression in poliomyelitis [121].

T cells of memory phenotype are predominant in T-cell leukemias including ALL, CLL, ATL and T-cell lymphomas [122, 123]. Two processes may be operative here also. Leukemogenesis may be more readily induced in memory cells or the memory phenotype may be acquired during neoplastic transformation.

Alterations in the proportions of circulating T cells of naive and memory phenotype have been extensively studied in HIV infection with complex and sometimes apparently contradictory findings. Part of the difficulty arises from the use of different markers to define memory cells and part from heterogeneity of patient population. However, several studies observe a consistent initial loss of CD4+ memory T cells [124, 125]. In asymptomatic seropositive donors, selective loss of memory cells and associated hyporesponsiveness to receptor-mediated activation can be observed. This is consistent with the finding that memory cells are the principal T-cell reservoir for HIV in infected individuals and are more readily infected in vitro with HIV than cells of naive phenotype [124, 126, 127]. Since memory cells may be in a low grade state of activation, this may provide an explanation for their vulnerability to infection.

It is clear that the exchange of lymphocytes between circulation and tissue is an extremely dynamic process with lymphocytes having a mean transit time in the circulation of approximately 30 min [128]. It is easy to envisage circumstances in which there is lymphopenia because a subset(s) adheres more efficiently to endothelial cells and is therefore relatively depleted from the circulating pool. This may be the mechanism of the apparent loss of circulating memory cells following rush sensitization with viper venom and in patients with leishmaniasis [129, 130]. In addition, it may explain why certain viral infections are often characterized by lymphopenia such as measles. The opposite effect can be seen in other viral infections, for example in infectious mononucleosis there is an increased percentage of circulating memory cells [131].

CD8 Naive and Memory Cells

Most studies of naive and memory cells have been done using unseparated T cells or CD4+ T cells. Although not as extensively studied, CD8 cells are thought to undergo a similar phenotypic transition from being CD45RA+ to CD45RO+, with CTL memory function confined to the cells of CD45RO+ phenotype [132–135]. Our unpublished studies of CD8 T cells in adult peripheral blood confirm the finding of others [136] that there are strikingly fewer memory cells and reciprocally many more naive cells among CD8+ cells when compared to CD4+ cells; this is observed not only in circulation but also in spleen, tonsil and lymph node [137]. This suggests that fewer CD8 cells have undergone antigen-specific priming than CD4+ cells. The explanation for this difference is not known. Functional data on CD8 naive and memory is limited [132–135, 138] though the properties of CD8 naive and memory cells appear to be similar to the better studied CD4 subsets.

Further Specialization among Memory Cells

Now that the general picture has been established of naive and memory cells, it is essential to go back and deal with some of the over- simplifications. We have discussed naive and memory cells as though they were homogeneous entities. In reality they are not.

Among the many molecules which are differentially expressed on naive and memory cells, we have already mentioned VLA-4α (CD49d) and VLA-β1 (CD29). CD45RA− memory cells generally express more of both molecules than do CD45RA+ naive cells. Although this is true, it is a major oversimplification. As Morimoto et al. [81] noted when they first described the VLA-β1 mAb 4B4, there is a generally reciprocal correlation between VLA-β1 (CD29/4B4) and CD45RA (2H4) in peripheral CD4+ T cells but that correlation is imperfect. Further analysis demonstrates that this generally reciprocal relationship breaks down completely when examining cells from tissues rather than circulating T cells [139, 140]. We find that a similar picture is seen with VLA-α4, -α5 and -α6; their expression is generally increased in memory cells but the correlation is not perfect [87, 139, 141].

Although it is tempting to shrug off these minor discordances, it is essential to overcome this impulse and seek an understanding of their significance in T-cell differentiation. When the molecules CD45RO, VLA-β1, VLA-α3, -α4, -α5 and -α6 are considered together, several simplifying generalizations emerge [139]. Among CD45RO− 'naive cells', there is very limited heterogeneity; almost all of the cells are low for VLA-β1 and virtually negative for VLA-α3, -α4, -α5 and -α6. In contrast, there is marked heterogeneity among the CD45RO-high memory cells. This heterogeneity is most striking when VLA-β1 and VLA-α4 are considered jointly (fig. 3). Naive cells are homogeneous in expression of these two markers; in contrast, these markers typically reveal three or more subpopulations among memory cells.

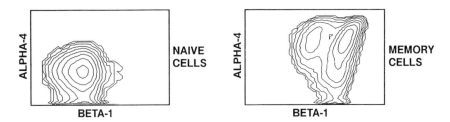

Fig. 3. Heterogeneity of memory but not naive cells for VLA-β1 and -α4. The left panel shows homogeneity of CD4$^+$ naive T cells gated for low expression of CD45RO. The right panel show heterogeneity of CD4$^+$ memory T cells gated for high expression of CD45RO.

The dominant population is high for VLA-β1 and expresses each of the α chains. However, up to 50% fall into other subsets with lower levels of either VLA-β1 or VLA-α4. The expression of VLA-α3, -α5 and -α6 also reveals heterogeneity, but this is virtually concordant with that detected by VLA-β1 and therefore adds no information to the analysis. These studies demonstrate that there is an extremely close correlation between the regulation of the proteins VLA-β1, VLA-α3, VLA-α5 and VLA-α6 on circulating CD4$^+$ T cells. In contrast, expression of VLA-β1 and VLA-α4 is not so perfectly correlated (fig. 3, memory cells). Implications of these finding are that: (1) the CD45RA/ RO isoform switch is regulated independently of the expression of VLA-β1 which in turn is regulated independently of the expression of VLA-α4; (2) at least in circulating CD4 cells, expression of VLA-β1 is regulated in a manner virtually concordant with VLA-α3, -α5 and -α6.

How Do We Interpret the Foregoing Results?

We propose that T-cell differentiation in vivo is complex and that many different pathways of differentiation occur following antigen exposure. The schema in figure 4 shows a homogeneous population of N = naive cells which can be stimulated by any of n different stimulation conditions (S1 through Sn). Differences between these stimulating conditions would undoubtedly include variables such as: anatomic site of stimulation, identity and activation state of the APC, concentration and nature of stimulating antigen, and cytokines present in the local microenvironment. The ensemble of these parameters constitute a unique stimulus designated 'Sn' and should induce a

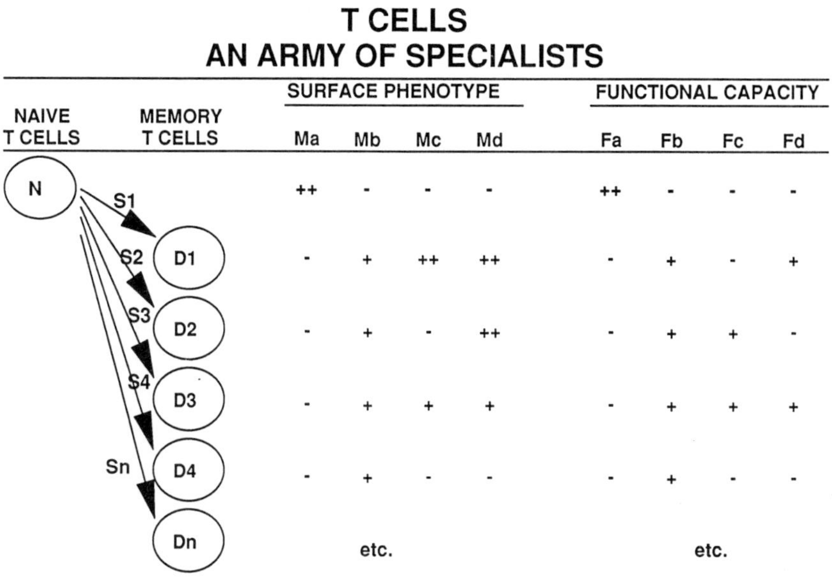

T CELLS
AN ARMY OF SPECIALISTS

NAIVE T CELLS	MEMORY T CELLS	SURFACE PHENOTYPE				FUNCTIONAL CAPACITY			
		Ma	Mb	Mc	Md	Fa	Fb	Fc	Fd
N (S1)		++	-	-	-	++	-	-	-
S2	D1	-	+	++	++	-	+	-	+
S3	D2	-	+	-	++	-	+	+	-
S4	D3	-	+	+	+	-	+	+	+
Sn	D4	-	+	-	-	-	+	-	-
	Dn			etc.				etc.	

Fig. 4. General schema of T-cell differentiation pathways. See text for explanation.

distinctive differentiation state of the memory cell indicated 'Dn'. Each such differentiation state would be expected to have a characteristic composite of phenotype and function. In the general schema shown in figure 4, a surface marker Ma would be present on all naive cells and be replaced by Mb on all memory cells. We think CD45RA and CD45RO are the closest real-life equivalents to these theoretical markers Ma and Mb. Other markers, for example Mc and Md, will subdivide memory cells. VLA-α4 and VLA-β1 are good candidates for such markers since their expression is augmented in some but not other CD45RO+ cells. The general schema in figure 4 predicts different functions for each differentiated subset of memory cells. We have not yet characterized the functional capacities of subsets of memory cells which are distinguished on the basis of differential expression of VLA-α4 and VLA-β1 (and its correlated α chains). We surmise that these subsets will differ markedly in their interactions with other cells and ECM. This is based on the known function of these integrins in interaction with a cell surface ligand on endothelium (VLA-4/VCAM-1) and with extracellular matrix molecules (fibronectin and laminin) [87, 142–145]. It is particularly attractive to postulate that differential homing will be observed since the integrin molecule LPAM-1 (which consists of VLA-α4 with a novel β chain β7 (originally βp)) has been reported to contribute to T-cell interactions with Peyer's patch HEV and possibly homing to lamina propria as well [146, 147]. Thus the expression of LPAM-1 by some circulating memory cells may identify a subset destined to recirculate through gut mucosal tissue.

In a similar vein, a number of other molecules appear to have important roles in defining memory cell heterogeneity. A particularly interesting molecule which identifies a subset of CD8[+] memory cells is HML-1 which is expressed by a small percentage of circulating lymphocytes and rarely expressed in lymphoid tissues such as spleen, mesenteric lymph node and tonsil, but is expressed by the vast majority of intestinal intraepithelial lymphocytes [102, 140]. This molecule, like VLA-4 and LPAM-1, is a member of the integrin family, apparently consisting of β7 and a potentially novel α chain [148]. It remains to be determined whether this molecule mediates very efficient homing of cells to the gut or whether its expression in gut T cells is a reflection of activation in the unique microenvironment of the gut [149]. Carbohydrate antigens also reveal functionally important heterogeneity on memory T cells. The carbohydrate known as cutaneous lymphocyte-associated antigen (CLA) detected by mAb HECA-452 is expressed by 10–20% of circulating CD4 cells [102]. This has been proposed as a ligand for the vascular adhesion molecule ELAM-1 which is expressed on inflamed skin endothelium and binds a subset of circulating memory cells which tend to migrate into skin [90, 150]. Another carbohydrate determinant whose expression subdivides memory T cells is CD60 [151, 152]; the significance of this heterogeneity is unclear at present.

One final marker of CD4[+] memory T-cell heterogeneity is the mAb PD7 which is specific for a CD45RB determinant [11, 14]. Given the extraordinary importance of CD45RA/CD45RO isoforms in our understanding of peripheral T-cell differentiation, we were excited by the prospect that the additional heterogeneity of CD4 cells revealed by another isoform would be equally revealing [153]. Although useful findings have emerged from our studies of CD45RB, CD45RB has not yet proved to be of such global importance as the CD45RA/CD45RO distinction. Our findings regarding CD45RB [14] can be summarized succinctly as follows:

(1) Although all CD4 cells express CD45RB, differences in the level of expression clearly subdivide memory (CD45RA[-]CD45RO[+]) cells; CD45RB is expressed at high levels on some (which we designate MEM1 cells) and at fourfold lower levels on others (which we designate MEM2). CD45RB is expressed at high levels on all naive cells (CD45RA[+]CD45RO[-]) (fig. 1).

(2) These findings regarding subdivision of memory cells pertain not only to T cells in circulation, but also to T cells in secondary lymphoid tissues. The absolute consistency with which this distinction is found in vivo convinces us that this distinction is important.

(3) Based on analysis of proliferative responses to recall antigens we find that both CD45RB-discriminated subsets of CD45RO⁺ cells include memory cells. Therefore, we use the terminology MEM1 and MEM2.

(4) Analysis of cytokines produced following activation of separated MEM1 and MEM2 cells with polyclonal activators fails to reveal differences between them in production of IL-2, IL-4 and IL-5, though MEM1 cells produce marginally more ($1.7\times$) IFN-γ than MEM2 cells.

(5) Despite a similar proliferative response of MEM1 and MEM2 cells to recall antigens, MEM1 cells consistently show enhanced proliferation following 24 h in culture.

(6) MEM1 and MEM2 cells are remarkably similar in the cell surface expression of a wide range of other molecules, including particularly the adhesion molecules whose expression differs between naive and memory cells.

(7) Regulation of CD45RB expression in vitro is more complex and reversible than CD45RA or CD45RO. Down-regulation of its expression can be observed when naive cells are activated in vitro, but so can apparent up-regulation in activated MEM2 cells, suggesting interconversion between MEM1 and MEM2 phenotypes.

Thus, our studies have not identified a dramatic functional difference between the MEM1 and MEM2 subsets. In part this may reflect the inherent problems with positive selection which is required for preparation of these subsets (since no reciprocal markers are available which allow negative selection of the same subsets). Nevertheless, we persist in the belief that CD45RB expression will prove to mark an important dichotomy between CD4 memory cells based on the pervasiveness of this distinction in vivo and the known importance of CD45.

In the last several paragraphs we have discussed findings of heterogeneity among memory (CD45RO⁺) cells from our studies and various recent published reports. Taken together, the identified phenotypic complexity among memory T cells seems formidable. The aggregate complexity of subsets will depend on whether these various markers fall into simple patterns of concordance/reciprocity or whether they are all relatively independent. So far, our studies reveal more evidence of independent relationships than simple ones, suggesting that with use of more markers even more complex heterogeneity will be detected. This leads to our view of memory T cells as an 'army of specialists'. Table 1 was formulated in a very general way because we believe that there will be a large number of distinct differentiation states. Although any subset may occur in quite low frequency in circulation, its capacity to home to a particular location or to interact with unique APC may confer on it a dominant role in particular situations. This concept of an army of specialists derived from the finding of phenotypic complexity complements much data from other species, demonstrating a marked tendency for antigen-primed T cells to recirculate to tissues where they were originally activated [154, 155]. This makes intuitive sense that, for example, a memory T cell specific for a respiratory pathogen would recirculate to the lung where that pathogen is likely to be re-encountered rather than to the gut or the skin where that is most unlikely. In contrast, there is data that naive cells are not restricted in their circulation to any anatomic site and may circulate throughout the body [155]. There is also data from human studies that the cells in tissue differ functionally from peripheral memory cells in ways not predictable from our present understanding of phenotypic subsets [156].

Consequently, it is clear that to fully grasp the subtleties of the physiology of the immune system, each of the markers and the subsets defined by their pattern of expression needs to be understood. Though enormous insight has been achieved, it is obvious that research on human CD4$^+$ T-cell subsets has only just begun.

Acknowledgements

We thank the many generous contributors of mAb, particularly the contributors to the 4th International Workshop of Leukocyte Differentiation Antigens; Gale Ginther Luce for outstanding technical assistance; the NIH blood bank and donors, and many colleagues for criticism and suggestions, particularly Tamas Schweighoffer, Yoji Shimizu and Gijs van Seventer.

References

1 Reinherz EL, Schlossman SF: The differentiation and function of human T lymphocytes. Cell 1980;19:821–827.
2 Sanders ME, Makgoba MW, Sharrow SO, Stephany D, Springer TA, Young HA, Shaw S: Human memory T lymphocytes express increased levels of three cell adhesion molecules (LFA-3, CD2, and LFA-1) and three other molecules (UCHL1,

CDw29, and Pgp-1) and have enhanced IFN-gamma production. J Immunol 1988; 140:1401–1407.

3 Tedder TF, Clement LT, Cooper MD: Human lymphocyte differentiation antigens HB-10 and HB-11. I. Ontogeny of antigen expression. J Immunol 1985;134:2983–2988.

4 Clement LT, Yamashita N, Martin AM: The functionally distinct subpopulations of human CD4+ helper/inducer T lymphocytes defined by anti-CD45R antibodies derive sequentially from a differentiation pathway that is regulated by activation-dependent post-thymic differentiation. J Immunol 1988;141:1464–1470.

5 Serra HM, Ledbetter JA, Krowka JF, Pilarski LM: Loss of CD45R (Lp220) represents a post-thymic T cell differentiation event. J Immunol 1988;140:1435–1441.

6 Williams AF, Barclay AN: Glycoprotein antigens of the lymphocyte surface and their purification of antibody affinity chromatography; in Weir DM, Harzenberg LA (eds): Handbook of Experimental Immunology. Oxford, Blackwell, 1985, pp 22.1–22.24.

7 Thomas ML: The leukocyte common antigen family. Annu Rev Immunol 1989;7: 339–369.

8 Hall LR, Streuli M, Schlossman SF, Saito H: Complete exon-intron organization of the human leukocyte common antigen (CD45) gene. J Immunol 1988;141:2781–2787.

9 Streuli M, Hall LR, Saga Y, Schlossman SF, Saito H: Differential usage of three exons generates at least five different mRNAs encoding human leukocyte common antigens. J Exp Med 1988;166:1548–1566.

10 Ralph SJ, Thomas ML, Morton CC, Trowbridge IS: Structural variants of human T200 glycoprotein (leukocyte-common antigen). EMBO J 1987;6:1251–1257.

11 Streuli M, Morimoto C, Schrieber M, Schlossman SF, Saito H: Characterization of CD45 and CD45R monoclonal antibodies using transfected mouse cell lines that express individual human leukocyte common antigens. J Immunol 1988;141:3910–3914.

12 Thomas ML, Lefrancois L: Differential expression of the leukocyte-common antigen family. Immunol Today 1988;9:320–326.

13 Sanders ME, Makgoba MW, Shaw S: Human naive and memory T cells: reinterpretation of helper-inducer and suppressor-inducer subsets. Immunol Today 1988;9:195–199.

14 Horgan KJ, Tanaka Y, van Seventer GA, Luce GEG, Nutman TB, Shaw S: CD45RB expression defines two interconvertible subsets of human CD4+ T cells with memory function. 1992, submitted.

15 Merkenschlager M, Terry L, Edwards R, Beverley PCL: Limiting dilution analysis of proliferative responses in human lymphocyte populations defined by the monoclonal antibody UCHL1: implications for differential CD45 expression in T cell memory formation. Eur J Immunol 1989;18:1653–1661.

16 Akbar AN, Terry L, Timms A, Beverley PC, Janossy G: Loss of CD45R and gain of UCHL1 reactivity is a feature of primed T cells. J Immunol 1988;140:2171–2178.

17 De Paoli P, Battistin S, Santini GF: Age-related changes in human lymphocyte subsets: progressive reduction of the CD4 CD45R (suppressor inducer) population. Clin Immunol Immunopathol 1988;48:290–296.

18 Hayward AR, Lee J, Beverley PCL: Ontogeny of expression of UCHL1 antigen on TcR-1+ (CD4/8) and TcRd+ T cells. Eur J Immunol 1989;19:771–773.

19 Koulova L, Yang SY, Dupont B: Identification of the anti-CD3-unresponsive subpopulation of CD4+, CD45RA+ peripheral T lymphocytes. J Immunol 1990;145: 2035–2043.

20 Byrne JA, Stankovic A, Cooper MD: Activated T cells in the peripheral lymphoid tissue of human fetuses. FASEB Presentation 1991;5:1702.

21 Pilarski LM, Gillitzer R, Zola H, Shortman K, Scollay R: Definition of the thymic generative lineage by selective expression of high molecular weight isoform of CD45 (T200). Eur J Immunol 1989;19:589–597.

22 Gillitzer R, Pilarski LM: In situ localization of CD45 isoforms in the human thymus indicates a medullary location for the thymic generative lineage. J Immunol 1990; 144:66–74.

23 Merkenschlager M, Fisher AG: CD45 isoform switching precedes the activation-driven death of human thymocytes by apoptosis. Int Immunol 1991;3:1–7.

24 Rothstein DM, Yamada A, Schlossman SF, Morimoto C: Cyclic regulation of CD45 isoform expression in a long-term human CD4$^+$CD45RA$^+$ T cell line. J Immunol 1991;146:1175–1183.

25 Serra HM, Ledbetter JA, Neil D, Pilarski LM: Apparent loss of 2H4$^+$ T cells in peripheral blood lymphocytes of normal donors: A 2H4$^-$ specific artefact unique to T cells. Hum Immunol 1988;23:281–288.

26 Schwinzer R, Wonigeit K: Genetically determined lack of CD45R$^-$ T cells in healthy individuals. Evidence for a regulatory polymorphism of CD45R antigen expression. J Exp Med 1990;171:1803–1808.

27 Poppema S, Lai R, Visser L: Antibody MT3 is reactive with a novel exon B-associated 190-kDa sialic acid-dependent epitope of the leukocyte common antigen complex. J Immunol 1991;147:218–223.

28 Lai R, Visser L, Poppema S: Tissue distribution of restricted leukocyte common antigens. Lab Invest 1991;64:844–854.

29 Bell EB, Sparshott SM: Interconversion of CD45R subsets of CD4 T cells in vivo. Nature 1990;348:163–166.

30 Sparshott SM, Bell EB, Sarawar SR: CD45R CD4 T cell subset-reconstituted nude rats: subset-dependent survival of recipients and bi-directional isoform switching. Eur J Immunol 1991;21:993–1000.

31 Shaw S, Luce GEG, Quinones R, Gress RE, Springer TA, Sanders ME: Two antigen-independent adhesion pathways used by human cytotoxic T cell clones. Nature 1986; 323:262–264.

32 Makgoba MW, Sanders ME, Luce GEG, Dustin ML, Springer TA, Clark EA, Mannoni P, Shaw S: ICAM-1, a ligand for LFA-1-dependent adhesion of B, T and myeloid cells. Nature 1988;331:86–88.

33 Shimizu Y, van Seventer GA, Horgan KJ, Shaw S: Roles of adhesion molecules in T cell recognition: Fundamental similarities between four integrins on resting human T cells (LFA-1, VLA-4, VLA-5, VLA-6) in expression, binding, and costimulation. Immunol Rev 1990;114:109–143.

34 Dang NH, Torimoto Y, Schlossman SF, Morimoto C: Human CD4 helper T cell activation: Functional involvement of two distinct collagen receptors, 1F7 and VLA integrin family. J Exp Med 1990;172:649–652.

35 Buckle A-M, Hogg N: Human memory T cells express intercellular adhesion molecule-1 which can be increased by interleukin-2 and interferon-gamma. Eur J Immunol 1990;20:337–341.

36 Akbar AN, Salmon M, Janossy G: The synergy between naive and memory cells during activation. Immunol Today 1991;12:184–188.

37 Akbar AN, Salmon M, Ivory K, Taki S, Pilling D, Janossy G: Human CD4$^+$CD45RO$^+$ and CD4$^+$CD45RA$^+$ T cells synergize in response to alloantigens. Eur J Immunol 1991;21:2517–2522.

38 Mackay CR, Marston WL, Dudler L: Naive and memory T cells show distinct pathways of lymphocyte recirculation. J Exp Med 1990;171:801–817.

39 Beverley PCL: Is T-cell memory maintained by cross-reactive stimulation? Immunol Today 1990;11:203–205.

40 Larsson EL, Anderson J, Coutinho A: Functional consequences of sheep red blood cell rosetting for human T-cells: gain of reactivity to mitogenic factors. Eur J Immunol 1978;8:693–696.

41 Wilkinson M, Morris AG: Role of the E receptor in interferon-gamma expression: Sheep erythrocytes augment interferon-gamma production by human lymphocytes. Cell Immunol 1984;86:109–117.

42 Ledbetter JA, Rabinovitch PS, Hellstrom I, Hellstrom KE, Grosmaire LS, June CH: Role of CD2 cross-linking in cytoplasmic calcium responses and T cell activation. Eur J Immunol 1988;18:1601–1608.

43 Breitmeyer JB, Faustman DL: Sheep erythrocyte rosetting induces multiple alterations in T lymphocyte function: inhibition of T cell receptor activity and stimulation of T11/CD2. Cell Immunol 1989;123:118–133.

44 Sanders ME, Makgoba MW, June CH, Young HA, Shaw S: Enhanced responsiveness of human memory T cells to CD2 and CD3 receptor-mediated activation. Eur J Immunol 1989;19:803–808.

45 Horgan KJ, van Seventer GA, Shimizu Y, Shaw S: Hyporesponsiveness of naive (CD45RA+) human T cells to multiple receptor-mediated stimuli but augmentation of responses by costimuli. Eur J Immunol 1990;20:1111–1118.

46 Byrne JA, Butler JL, Cooper MD: Differential activation requirements for virgin and memory T cells. J Immunol 1988;141:3249–3257.

47 Byrne JA, Butler JL, Reinherz EL, Cooper MD: Virgin and memory T cells have different requirements for activation via the CD2 molecule. Int Immunol 1989;1:29–35.

48 Schraven B, Roux M, Hutmacher B, Meuer SC: Triggering of the alternative pathway of human T cell activation involves members of the T200 family of glycoproteins. Eur J Immunol 1989;19:397–403.

49 Alexander DR, Cantrell DA: Kinase and phosphatases in T-cell activation. Immunol Today 1989;10:200–205.

50 Pingel JT, Thomas ML: Evidence that the leukocyte-common antigen is required for antigen-induced T lymphocyte proliferation. Cell 1989;58:1055–1065.

51 Koretzky GA, Picus J, Schultz T, Weiss A: Tyrosine phosphatase CD45 is required for T cell antigen receptor and CD2-mediated activation of a protein tyrosine kinase and interleukin-2 production. Proc Natl Acad Sci USA 1991;88:2037–2041.

52 Rudd CE, Anderson P, Morimoto C, Streuli M, Schlossman SF: Molecular interactions, T-cell subsets and a role of the CD4/CD8:56lck complex in human T-cell activation. Immunol Res 1989;111:225–266.

53 Mustelin T, Coggeshall KM, Altman A: Rapid activation of the T-cell tyrosine protein kinase pp56lck by the CD45 phosphotyrosine phosphatase. Proc Natl Acad Sci USA 1989;86:6302–6306.

54 Ostergaard HL, Trowbridge IS: Coclustering CD45 with CD4 or CD8 alters the phosphorylation and kinase activity of p56lck. J Exp Med 1990;172:347–350.

55 Dianzani U, Luqman M, Rojo J, Yagi J, Baron JL, Woods A, Janeway CA, Bottomly K: Molecular associations on the T cell surface correlate with immunological memory. Eur J Immunol 1990;20:2249–2257.

56 Stamenkovic I, Sgroi D, Aruffo A, Sun Sy M, Anderson T: The B lymphocyte adhesion molecule CD22 interacts with leukocyte common antigen CD45RO on T cells and a2–6 sialyltransferase, CD75, on B cells. Cell 1991;66:1133–1144.

57 Ledbetter JA, Tonks NK, Fischer EH, Clark EA: CD45 regulates signal transduction and lymphocyte activation by specific association with receptor molecules on T or B cells. Proc Natl Acad Sci USA 1988;85:8628–8632.

58 Schraven B, Samstag Y, Altevogt P, Meuer SC: Association of CD2 and CD45 on human T lymphocytes. Nature 1990;345:71–74.

59 Kiener PA, Mittler RA: CD45-protein tyrosine phosphatase cross-linking inhibits T cell receptor CD3-mediated activation in human T cells. J Immunol 1989;143:23–28.

60 Deans JP, Shaw J, Pearse MJ, Pilarski LM: CD45R as a primary signal transducer stimulating IL-2 and IL-2R mRNA synthesis by CD3⁻4⁻8⁻ thymocytes. J Immunol 1989;143:2425–2430.

61 Dohlsten M, Hedlund G, Sjögren HO, Carlsson R: Two subsets of human CD4⁺ T helper cells differing in kinetics and capacities to produce interleukin-2 and interferon-gamma can be defined by the Leu-18 and UCHL1 monoclonal antibodies. Eur J Immunol 1988;18:1173–1178.

62 Hedlund G, Dohlsten M, Sjögren HO, Carlsson R: Maximal interferon-gamma production and early synthesis of interleukin-2 by CD4⁺CDw29⁻CD45R⁻p80⁻ human T lymphocytes. Immunology 1989;66:49–53.

63 Salmon M, Kitas GD, Bacon PA: Production of lymphokine mRNA by CD45R⁺ and CD45R⁻ helper T cells from human peripheral blood and by human CD4⁺ T cell clones. J Immunol 1989;143:907–912.

64 Salmon M, Kitas GD, Hill Gaston JS, Bacon PA: Interleukin-2 production and response by helper T-cell subsets in man. Immunology 1988;65:81–85.

65 Fischer H, Dohlsten M, Andersson U, Hedlund G, Ericsson P, Hansson J, Sjögren HO: Production of TNF-α and TNF-β by staphylococcal enterotoxin A activated human T cells. J Immunol 1990;144:4663–4669.

66 Zutter MM, Santoro SA: Widespread histologic distribution of the alpha-2-beta-1-integrin cell-surface collagen receptor. Am J Pathol 1990;137:113–120.

67 Mosmann TR, Coffman RL: Heterogeneity of cytokine secretion patterns and functions of helper T cells. Adv Immunol 1989;46:111–147.

68 Street NE, Mosmann TR: Functional diversity of T lymphocytes due to secretion of different cytokine patterns. FASEB J 1991;5:171–176.

69 Parronchi P, Macchia D, Piccinni M-P, Biswas P, Simonelli C, Maggi E, Ricci M, Ansari AA, Romagnani S: Allergen- and bacterial antigen-specific T-cell clones established from atopic donors show a different profile of cytokine production. Proc Natl Acad Sci USA 1991;88:4538–4542.

70 Romagnani S: Human Th1 and Th2 subsets: doubt no more. Immunol Today 1991; 12:256–257.

71 Luqman M, Johnson P, Trowbridge I, Bottomly K: Differential expression of the alternatively spliced exons of murine CD45 in Th1 and Th2 cell clones. Eur J Immunol 1991;21:17–22.

72 Haanen JB, De Waal Malefijt R, Res PC, Kraakman EM, Ottenhoff TH, de Vries RR, Spits H: Selection of a human T helper type 1-like T cell subset by mycobacteria. J Exp Med 1991;174:583–592.

73 Yssel H, Shanafelt MC, Soderberg C, Schneider PV, Anzola J, Peltz G: *Borrelia*

burgdorferi activates a T helper type 1-like T cell subset in Lyme arthritis. J Exp Med 1991;174:593–601.

74 Damle NK, Doyle LV: Stimulation via the CD3 and CD28 molecules induces responsiveness to IL-4 in CD4+CD29+CD45R⁻ memory T lymphocytes. J Immunol 1989;143:1761–1767.

75 Wasik MA, Morimoto C: Differential effects of cytokines on proliferative response of human CD4+ T lymphocyte subsets stimulated via T cell receptor-CD3 complex. J Immunol 1990;144:3334–3340.

76 Kasahara Y, Miyawaki T, Kato K, Kanegane H, Yachie A, Yokoi T, Taniguchi N: Role of interleukin-6 for differential responsiveness of naive and memory CD4+ T cells in CD2-mediated activation. J Exp Med 1990;172:1419–1424.

77 Gaya A, Alsinet E, Martorell J, Places L, Calle Odl, Yague J, Vives J: Inhibitory effect of IL-4 on the sepharose-CD3-induced proliferation of the CD4CD45RO human T cell subset. Int Immunol 1990;2:685–689.

78 Schall TJ, Bacon K, Toy KJ, Goeddel DV: Selective attraction of monocytes and T lymphocytes of the memory phenotype by cytokine RANTES. Nature 1990;347: 669–671.

79 Panzer S, Geller RL, Bach FH: Purified human T cells stimulated with crosslinked anti-CD3 monoclonal antibody OKT3: rIL-1 is a costimulatory factor for CD4+CD29+CD45RA⁻ T cells. Scand J Immunol 1990;32:359–371.

80 Morimoto C, Letvin NL, Distaso JA, Aldrich WR, Schlossman SF: The isolation and characterization of the human suppressor inducer T cell subset. J Immunol 1985; 134:1508–1515.

81 Morimoto C, Letvin NL, Boyd AW, Hagan M, Brown HM, Kornacki MM, Schlossman SF: The isolation and characterization of the human helper inducer T cell subset. J Immunol 1985;134:3762–3769.

82 Hirohata S, Lipsky PE: T cell regulation of human B cell proliferation and differentiation. Regulatory influences of CD45R+ and CD45R⁻ T4 cell subsets. J Immunol 1989;142:2597–2607.

83 Sleasman JW, Henderson M, Barrett DJ: Con A-induced suppressor cell function depends on the activation of the CD4+CD45RA inducer T cell subpopulation. Cell Immunol 1991;133:367–378.

84 Gerli R, Bertotto A, Crupi S, Arcangeli C, Marinelli I, Spinozzi F, Cernetti C, Angelella P, Rambotti P: Activation of cord T lymphocytes. I. Evidence of a defective T cell mitogenesis induced through the CD2 molecule. J Immunol 1989;142:2583–2589.

85 Ehlers S, Smith KA: Differentiation of T cell lymphokine gene expression: the in vitro acquisition of T cell memory. J Exp Med 1991;173:25–36.

86 Lewis DB, Yu CC, Meyer J, English BK, Kahn SJ, Wilson CB: Cellular and molecular mechanisms for reduced interleukin-4 and interferon-gamma production by neonatal T cells. J Clin Invest 1991;87:194–202.

87 Shimizu Y, van Seventer GA, Horgan KJ, Shaw S: Regulated expression and function of three VLA (beta-1) integrin receptors on T cells. Nature 1990;345:250–253.

88 Shimizu Y, Newman W, Graber N, Horgan KJ, Beall LD, Gopal TV, van Seventer GA, Shaw S: Four molecular pathways of T cell adhesion to endothelial cells: roles of LFA-1, VCAM-1 and ELAM-1 and changes in pathway hierarchy under different activation conditions. J Cell Biol 1991;113:1203–1212.

89 Damle NK, Doyle LV: Ability of human T lymphocytes to adhere to vascular endothelial cells and to augment endothelial permeability to macromolecules is linked to their state of post-thymic maturation. J Immunol 1990;144:1233–1240.

90 Shimizu Y, Shaw S, Graber N, Gopal TV, Horgan KJ, van Seventer GA, Newman W: Activation-independent binding of human memory T cells to ELAM-1. Nature 1991; 349:799–802.

91 Bos JD, Hagenaars C, Das PK, Krieg SR, Voorn WJ, Kapsenberg ML: Predominance of memory T-cells (CD4+, CDw29+) over naive T-cells (CD4+, CD45R+) in both normal and diseased human skin. Arch Dermatol Res 1989;281:24–30.

92 Saltini C, Kirby M, Trapnell BC, Tamura N, Crystal RG: Biased accumulation of T lymphocytes with 'memory'-type CD45 leukocyte common antigen gene expression on the epithelial surface of the human lung. J Exp Med 1990;171:1123–1140.

93 James SP, Fiocchi C, Graeff AS, Strober W: Phenotypic analysis of lamina propria lymphocytes. Predominance of helper-inducer and cytolytic T cell phenotypes and deficiency of suppressor-inducer phenotypes in Crohn's disease and control patients. Gastroenterology 1986;91:1483–1489.

94 Brandtzaeg P, Bosnes V, Halstensen TS, Scott H, Sollid LM, Valnes KN: T lymphocytes in human gut epithelium preferentially express the alpha/beta antigen receptor and are often CD45/UCHL1-positive. Scand J Immunol 1989;30:123–128.

95 Pitzalis C, Kingsley GH, Covelli M, Meliconi R, Markey A, Panayi GS: Selective migration of the human helper-inducer memory T cell subset: confirmation by in vivo cellular kinetic studies. Eur J Immunol 1991;21:369–376.

96 Bertotto A, Gerli R, Fabietti G, Crupi S, Arcangeli C, Scalise F, Vaccaro R: Human breast milk T lymphocytes display the phenotype and functional characteristics of memory T cells. Eur J Immunol 1990;20:1877–1880.

97 Horgan KJ, Shaw S: Immunological memory; in Roitt IM (ed): Encyclopedia of Immunology. London, Sanders Scientific Publications, 1990.

98 Kupper TS: Immune and inflammatory processes in cutaneous tissues: Mechanisms and speculations. J Clin Invest 1990;86:1783–1789.

99 Bogen SA, Weinberg DS, Abbas AK: Histologic analysis of T lymphocyte activation in reactive lymph nodes. J Immunol 1991;147:1537–1541.

100 Janossy G, Bofill M, Rowe D, Muir J, Beverley PC: The tissue distribution of T lymphocytes expressing different CD45 polypeptides. Immunology 1989;66:517–525.

101 Tedder TF, Matsuyama T, Rothstein D, Schlossman SF, Morimoto C: Human antigen-specific memory T cells express the homing receptor (LAM-1) necessary for lymphocyte recirculation. Eur J Immunol 1990;20:1357–1366.

102 Picker LJ, Terstappen LWMM, Rott LS, Streeter PR, Stein H, Butcher EC: Differential expression of homing-associated adhesion molecules by T cell subsets in man. J Immunol 1990;145:3247–3255.

103 Akbar AN, Amlot PL, Ivory K, Timms A, Janossy G: Inhibition of alloresponsive naive and memory T cells by CD7 and CD25 antibodies and by cyclosporine. Transplantation 1990;50:823–829.

104 Shimizu Y, van Seventer GA, Horgan KJ, Shaw S: T cell-specific accessory molecules CD7 and CD28 modulate integrin-mediated T cell adhesion. 1992;175:577–582.

105 Tanaka Y, Albelda SM, Horgan KJ, van Seventer GA, Shimizu Y, Buck CA, Shaw S: CD31/PECAM-1 is on unique T cells and functions in an adhesion cascade. Submitted.

106 Shimizu Y, Newan W, Tanaka Y, Shaw S: Lymphocyte/endothelial interactions. Immunol Today, in press.

107 Modlin RL, Melancon-Kaplan J, Young SM, Pirmez C, Kino H, Convit J, Rea TH, Bloom BR: Learning from lesions: patterns of tissue inflammation in leprosy. Proc Natl Acad Sci USA 1988;85:1213–1217.

108 Barnes PF, Mistry SD, Cooper CL, Pirmez C, Rea TH, Modlin RL: Compartmentalization of a CD4⁺ T lymphocyte subpopulation in tuberculous pleuritis. J Immunol 1989;142:1114–1119.

109 Kingsley G, Pitzalis C, Kyriazis N, Panayi GS: Abnormal helper-inducer/suppressor-inducer T-cell subset distribution and T-cell activation status are common to all types of chronic synovitis. Scand J Immunol 1988;28:225–232.

110 Pitzalis C, Kingsley G, Haskard D, Panayi G: The preferential accumulation of helper-inducer T lymphocytes in inflammatory lesions: evidence for regulation by selective endothelial and homotypic adhesion. Eur J Immunol 1988;18:1397–1404.

111 Chofflon M, Gonzalez V, Weiner HL, Hafler DA: Inflammatory cerebrospinal fluid T cells have activation requirements characteristic of CD4⁺CD45RA⁻ T cells. Eur J Immunol 1989;19:1791–1795.

112 Lasky HP, Bauer K, Pope RM: Increased helper inducer and decreased suppressor inducer phenotypes in the rheumatoid joint. Arthritis Rheum 1988;31:52–59.

113 Miescher S, Schreyer M, Barras C, Capasso P, Von Fliedner V: Sparse distribution of gamma/delta T lymphocytes around human epithelial tumors predominantly infiltrated by primed/memory T cells. Cancer Immunol Immunother 1990;32:81–87.

114 Sato K, Miyasaka N, Yamaoka K, Okuda M, Yata J, Nishioka K: Quantitative defect of CD4⁺2H4⁺ cells in systemic lupus erythematosus and Sjögren's syndrome. Arthritis Rheum 1987;30:1407–1411.

115 Morimoto C, Steinberg AD, Letvin NL, Hagan M, Takeuchi T, Daley J, Levine H, Schlossman SF: A defect of immunoregulatory T cell subsets in systemic lupus erythematosus patients demonstrated with anti-2H4 antibody. J Clin Invest 1987;79:762–768.

116 Morimoto C, Hafler DA, Weiner HL, Letvin NL, Hagan M, Daley J, Schlossman SF: Selective loss of the suppressor-inducer T-cell subset in progressive multiple sclerosis. N Engl J Med 1987;316:67–72.

117 Rose LM, Ginsburg AH, Rothstein TL, Ledbetter JA, Clark EA: Selective loss of a subset of T-helper cells in active multiple sclerosis. Proc Natl Acad Sci USA 1985;82:7389–7393.

118 Sanders ME, Makgoba MW, Shaw S: Alterations in T cell subsets in multiple sclerosis and other autoimmune diseases. Lancet 1988;ii:1021.

119 Mylvaganam R, Ahn YS, Sprinz PG, Garcia RO, Harrington WJ: Sex difference in the CD4⁺CD45R⁺ T lymphocytes in normal individuals and its elective decrease in women with idiopathic thrombocytopenic purpura. Clin Immunol Immunopathol 1989;52:473–485.

120 Rose LM, Ginsberg AH, Rothstein TL, Ledbetter JA, Clark EA: Fluctuations of CD4⁺ T-cell subsets in remitting-relapsing multiple sclerosis. Ann Neurol 1988;24:192–199.

121 Ginsberg AH, Gale MJ, Rose LM, Clark EA: T-cell alterations in late postpoliomyelitis. Arch Neurol 1989;46:497–501.

122 Yamada Y, Ichimaru M, Shiku H: Adult T cell leukemia cells are CD4⁺ CDw29⁺ T cell origin and secrete a B cell differentiation factor. Br J Haematol 1989;72:370–377.

123 Worner I, Matutes E, Beverley PCL, Catovsky D: The distribution of CD45R, CD29 and CD45RO (UCHL1) antigens in mature CD4-positive T-cell leukaemias. Br J Haematol 1990;74:439–444.

124 Schnittman SM, Lane HC, Greenhouse J, Justement JS, Baseler M, Fauci AS: Preferential infection of CD4⁺ memory T cells by human immunodeficiency virus type 1: evidence for a role in the selective T-cell functional defects observed in infected individuals. Proc Natl Acad Aci USA 1990;87:6058–6062.

125 Van Noesel CJM, Gruters RA, Terpstra FG, Schellekens PTA, Van Lier RAW, Miedema F: Functional and phenotypic evidence for a selective loss of memory T cells in asymptomatic HIV-infected men. J Clin Invest 1990;86:293–299.

126 Wood GS, Burns BF, Dorfman RF, Warnke RA: In situ quantitation of lymph node helper, suppressor, and cytotoxic T cell subsets in AIDS. Blood 1986;67: 596–603.

127 Gallatin WM, Gale MJ, Hoffman PA, Willerford DM, Draves KE, Benveniste RE, Morton WR, Clark EA: Selective replication of simian immunodeficiency virus in a subset of CD4⁺ lymphocytes. Proc Natl Acad Sci USA 1989;86:3301–3305.

128 Schick P, Trepel F, Eder M, Matzner M, Benedek S, Theml H, Kaboth W, Begemann H, Fliedner TM: Autotransfusion of ³H-cytidine-labelled blood lymphocytes in patients with Hodgkin's disease and non-Hodgkin patients. Acta Haematol 1975;53: 206–218.

129 Tilmant L, Dessaint JP, Tsicopoulos A, Tonnel AB, Capron A: Concomitant augmentation of CD4⁺ CD45R⁺ suppressor inducer subset and diminution of CD4⁺ CDw29⁺ helper inducer subset during rush hyposensitization in hymenoptera venom allergy. Clin Exp Immunol 1989;76:13–18.

130 Cillari E, Milano S, Dieli M, Maltese E, Di Rosa S, Mansueto S, Salerno A, Liew FY: Reduction in the number of UCHL-1⁺ cells and IL-2 production in the peripheral blood of patients with visceral leishmaniasis. J Immunol 1991;146:1026–1030.

131 Miyawaki T, Kasahara Y, Kanegane H, Ohta K, Yokoi T, Yachie A, Taniguchi N: Expression of CD45RO (UCHL1) by CD4⁺ and CD8⁺ T cells as a sign of in vivo activation in infectious mononucleosis. Clin Exp Immunol 1991;83:447–451.

132 Yamashita N, Clement LT: Phenotypic characterization of the post-thymic differentiation of human alloantigen-specific CD8⁺ cytotoxic T lymphocytes. J Immunol 1989;143:1518–1523.

133 Beverley PC: Immunological memory in T cells. Curr Opin Immunol 1991;3:355–360.

134 Merkenschlager M, Beverley PCL: Evidence for differential expression of CD45 isoforms by precursors for memory-dependent and independent cytotoxic responses: human CD8 memory CTLp selectively express CD45RO (UCHL1). Int Immunol 1989;1:450–459.

135 Akbar AN, Amlot PL, Timms A, Lombardi G, Lechler R, Janossy G: The development of primed/memory CD8⁺ lymphocytes in vitro and in rejecting kidneys after transplantation. Clin Exp Immunol 1990;81:225–231.

136 Richards SJ, Jones RA, Roberts BE, Patel D, Scott CS: Relationship between 2H4 (CD45RA) and UCHL1 (CD45RO) expression by normal blood CD4⁺CD8⁻, CD4⁻CD8⁺, CD4⁻CD8dim⁺, CD3⁺CD4⁻CD8⁻ and CD3⁻CD4⁻CD8⁻ lymphocytes. Clin Exp Immunol 1990;81:149–155.

137 Tanaka Y, Horgan KJ, Luce GE, Shaw S: Phenotypic heterogeneity of CD8 cells. In preparation 1992.

138 Schlunck T, Schraut W, Riethmuller G, Loms Ziegler-Heitbrock HW: Inverse relationship of Ca²⁺ mobilization and cell proliferation in CD8⁺ memory and virgin T cells. Eur J Immunol 1990;20:1957–1963.

139 Horgan KJ, Tanaka Y, Luce GG, Schweighoffer T, Shimizu Y, Shaw S: VLA-4, VLA-beta-1 and CD45RO each mark independent differentiation status of circulating human T cells. In preparation 1991.

140 Jarry A, Cerf-Bensussan N, Brousse N, Selz F, Guy-Grand D: Subsets of CD3⁺ (T cell receptor alpha/beta or gamma/delta) and CD3⁻ lymphocytes isolated from normal human gut epithelium display phenotypical features different from their counterparts in peripheral blood. Eur J Immunol 1990;20:1097–1103.

141 Shimizu Y, van Seventer GA, Horgan KJ, Shaw S: Costimulation of proliferative responses of resting CD4⁺ T cells by the interaction of VLA-4 and VLA-5 with fibronectin or VLA-6 with laminin. J Immunol 1990;145:59–67.

142 Wayner EA, Garcia-Pardo A, Humphries MJ, McDonald JA, Carter WG: Identification and characterization of the T lymphocyte adhesion receptor for an alternative cell attachment domain (CS-1) in plasma fibronectin. J Cell Biol 1989;109:1321–1330.

143 Guan JL, Hynes RO: Lymphoid cells recognize an alternatively spliced segment of fibronectin via the integrin receptor alpha-4-beta-1. Cell 1990;60:53–61.

144 Garcia-Pardo A, Wayner EA, Carter WG, Ferreira OC Jr: Human B lymphocytes define an alternative mechanism of adhesion to fibronectin. The interaction of the α4β1 integrin with the LHGPEILDVPST sequence of the type III connecting segment is sufficient to promote cell attachment. J Immunol 1990;144:3361–3366.

145 Shimizu Y, Shaw S: Lymphocyte interactions with extracellular matrix. FASEB J 1991;5:2292–2299.

146 Holzmann B, McIntyre BW, Weissman IL: Identification of a murine Peyer's patch-specific lymphocyte homing receptor as an integrin molecule with an alpha chain homologous to human VLA-4. Cell 1989;56:37–46.

147 Holzmann B, Weissman IL: Peyer's patch-specific lymphocyte homing receptors consist of a VLA-4-like alpha chain associated with either of two integrin beta chains, one of which is novel. EMBO J 1989;8:1735–1741.

148 Yuan Q, Jiang W-M, Hollander D, Leung E, Watson JD, Krissansen GW: Identity between the novel integrin β₇ subunit and an antigen found highly expressed on intraepithelial lymphocytes in the small intestine. Biochem Biophys Res Commun 1991;176:1443–1449.

149 Schieferdecker HL, Ullrich R, Weiss-Breckwoldt AN, Schwarting R, Stein H, Riecken E-O, Zeitz M: The HML-1 antigen of intestinal lymphocytes is an activation antigen. J Immunol 1990;144:2541–2549.

150 Picker LJ, Kishimoto TK, Smith CW, Warnock RA, Butcher EC: ELAM-1 is an adhesion molecule for skin-homing T cells. Nature 1991;349:796–799.

151 Higgs JB, Zeldes W, Kozarsky K, Schteingart M, Kan L, Bohlke P, Krieger K, Davis W, Fox DA: A novel pathway of human T lymphocyte activation: Identification by a monoclonal antibody generated against a rheumatoid synovial T cell line. J Immunol 1988;140:3758–3765.

152 Fox DA, Millard JA, Kan L, Zeldes WS, Davis W, Higgs J, Emmrich F, Kinne RW: Activation pathways of synovial T lymphocytes; expression and function of the UM4D4/CDw60 antigen. J Clin Invest 1990;86:1124–1136.

153 Mason D, Powrie F: Memory CD4$^+$ T cells in man from two subpopulations, defined by their expression of isoforms of the leukocyte common antigen, CD45. Immunology 1990;70:427–433.

154 Butcher EC: The regulation of lymphocyte traffic. Curr Top Microbiol Immunol 1986;128:85–122.

155 Mackay CR: T-cell memory: the connection between function, phenotype and migration pathway. Immunol Today 1991;12:189–192.

156 Pirzer UC, Schurman G, Post S, Betzler M, Meuer SC: Differential responsiveness to CD3-Ti vs. CD2-dependent activation of human intestinal T lymphocytes. Eur J Immunol 1990;20:2339–2342.

Stephen Shaw, NlH Bldg 10, Room 4B17, Bethesda, MD 20892 (USA)

Coffman RL (ed): Regulation and Functional Significance of T-Cell Subsets.
Chem Immunol. Basel, Karger, 1992, vol 54, pp 103–116

Murine CD4+ T-Cell Subsets Generated by Antigen-Independent and Dependent Mechanisms

Kyoko Hayakawa, Betsy T. Lin, Richard R. Hardy

Institute for Cancer Research, Fox Chase Cancer Center, Philadelphia, Pa., USA

Introduction

An intriguing recent finding in immunology is the delineation of functional heterogeneity in CD4+ T cells. In addition to the conventional role of CD4+ T cells as helper cells [1, 2], they have also been shown to exhibit a suppressive function. Key to this initial work has been the association of these functional differences with distinct fractions of CD4+ T cells [3, 4]. The heterogeneity of CD4+ T-cell function was further revealed by work from a different direction. A series of studies on murine CD4+ T-cell lines demonstrated the presence of two types of lines, each possessing the potential to secrete distinct cytokines, denoted as Th1 and Th2 in mice [5]. These data have provided initial evidence to show that CD4+ T cells become restricted to subsets with distinct effector functions [6, 7].

The demonstration of both functional and phenotypic heterogeneity has attracted considerable attention in the study of CD4+ T cells, cells which play an important role in a variety of immune responses [8]. It has led to studies aimed at understanding how functionally heterogeneous T-cell subsets are generated during the process of T-cell differentiation and maturation. One clear observation, found in several species, is that the majority of CD4+ T cells in the periphery secrete predominantly IL-2 upon stimulation, whereas a small fraction of cells expressing a unique cell phenotype can secrete a more diverse spectrum of cytokines with less production of IL-2 [9–13]. Cells phenotypically and functionally similar to this small subset could be induced in vitro or in vivo following antigenic stimulation [14–17]. These data strongly suggest that stimulation by antigen results in altered regulation of cytokine production, providing an important concept, that is, maturation of T cells into memory cells [9, 18]. This leads one to consider that at least some of the heterogeneity of CD4$^+$ T cells found in laboratory animals may be accounted for by the generation of memory T cells by stimulation encountered in their normal environment.

Consequently, many questions have been raised as to the mechanisms to generate functional heterogeneity in CD4$^+$ T cells. For example, it is not clear whether the commitment to restricted cytokine potential (Th1 or Th2) is an antigen-dependent process or not [6, 8, 19], since this phenotype cannot be explained simply on the basis of naive versus memory cell differentiation stage. Rather, it could be argued that such different funtional CD4$^+$ T cells might be generated independent of antigen stimulation during a normal differentiation process leading to distinct subsets. Alternatively, if their origin is antigen-dependent, then the mechanism of regulation in the maturation process into memory cells becomes an important issue for further study [6]. Our strategy to answer these questions has been first to delineate the CD4$^+$ T-cell differentiation pathway from the thymus to the periphery and then to examine whether the cell fractions with diverse cytokine potential (apart from IL-2 alone) are all generated as a result of antigenic selection. We introduce here three secondary cell subsets present in the thymus or spleen of mice. Each subset is distinguished by its characteristic responsiveness but all share the ability to secrete a diverse array of cytokines. The evidence we present indicates that they are all generated by antigenic selection in the natural environment, supporting the concept of the importance of antigen in altering cytokine potential.

Murine CD4$^+$ T-Cell Subsets in the Thymus and Periphery Defined by 6C10 and 3G11 Expression

Several cell surface antigens have been shown to be altered depending on T-cell differentiation and maturation stage. We introduce here the 6C10 and 3G11 antigens. They are detected by two anti-T-cell autoantibodies SM6C10 and SM3G11, respectively [13]. Expression of both on the cell surface is restricted to the T-cell lineage among hematopoietic cells. 6C10 expression is dependent on cell surface Thy-1 expression, and probably is a carbohydrate epitope on the Thy-1 glycoprotein itself [13, 20]. 3G11 is a sialated carbohydrate antigen expressed on a unique ganglioside(s) [J.M. Greer et al.,

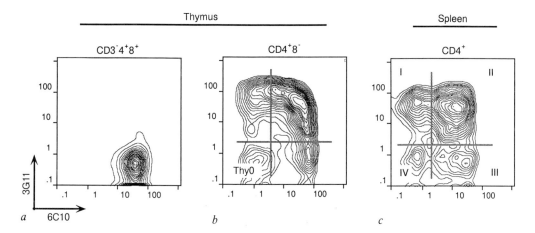

Fig. 1. a–c CD4+ T-cell subsets defined by 6C10 and 3G11 levels in the thymus and spleen. Thymocytes and spleen cells of 2 mo. old BALB/c mouse were used for multicolor immunofluorescence analysis by flow cytometry. Cells were simultaneously stained with antibodies to the molecules indicated in addition to CD8. 6C10/3G11 distribution on (CD8CD4)+CD3− cells and CD4+CD8− cells is shown. Lines in the figures demarcate the subsets. ThyO is the thymic 6C10−(HSA−)3G11−CD4+ subset, and Fr.I–IV are the splenic CD4+ subsets.

submitted]. As figure 1a shows, similar to Thy-1, 6C10 is highly expressed on CD3−CD4+CD8+ immature cells in the thymus [13, 21]. Although expression of 6C10 decreases on the CD4+(CD8−) single positive cells in the thymus (referred to as thymic CD4+ T cells hereafter), many cells retain detectable levels of 6C10 whereas some are truly 6C10− (fig. 1b). Unlike 6C10, 3G11 expression is not found (or is very low) on the immature cells but is highly expressed on CD3+ T cells in the thymus, including both CD4+CD8+ and CD4+ (or CD8+) single positive cells (fig. 1b). As figure 1b shows, the majority of CD4+ T cells in the thymus express 3G11 and two CD3+ cell fractions can be resolved as 6C10+3G11+ and 6C10−3G11+. In addition, as figure 1b also reveals, a minor 6C10−3G11−(CD3+) cell subset is found in the thymic CD4+ T cells (described below). Different from these three cell subsets, 6C10+3G11− cells in the thymus are found to be CD3−. In spleen (and lymph nodes) of adult mice, four CD4+ T-cell subsets (Fr. I–IV) are also definable by 6C10 and 3G11 (fig. 1c). Similar to the thymic CD4+ T cells, 3G11+ cells (Fr.I and II) comprise the majority of peripheral CD4+ T cells, and 3G11− cells (Fr. III and IV) are minor, but both are CD3+.

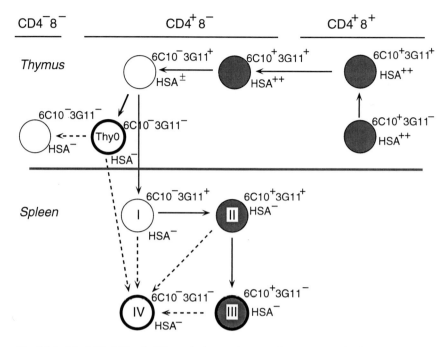

Fig. 2. Model of CD4+ T-cell differentiation and maturation pathways in the thymus and the peripheral lymphoid organs. Changes of 6C10 and 3G11 levels (and HSA) are described here. Dashed lines indicate provisional pathways whose definition requires further studies. 6C10+ cell stages are marked in gray to stress that 6C10 is re-expressed in the periphery. The three types of secondary cell subsets are marked by thick circles.

Cell Differentiation and Cell Surface Molecule Expression:
Antigen-Independent Progression in the Thymus and Periphery

Figure 2 summarizes our current understanding of CD4+ T-cell differentiation and maturation pathways in the thymus and periphery, based on the subsets defined by 6C10 and 3G11 expression.

Thymus

After selection into the CD4+ T-cell subset in the thymus, cells progress further acquiring functional competence. In the thymus, this progression has been noted by the down-regulation of heat-stable antigen (HSA) expression [22, 23]. HSA level is high on the immature T-lineage cells in the thymus, lower on thymic CD4+ T cells, and absent from peripheral T cells [22, 24]. We found that 6C10/Thy-1 expression is also decreased coordinately with cell differentiation in the thymus as with HSA [Hayakawa et al., submitted]. 6C10+3G11+ cells in the thymic CD4+ cell fraction shown in figure 1b are all HSA+, and 6C10-3G11+ cells include both HSA^low and HSA- subsets. 6C10-3G11- cells are all HSA-. Functional progression was demonstrated by the fact that only 6C10-3G11+ cells respond to anti-CD3 stimulation despite the fact that both 6C10+ and 6C10-3G11+ cells express similar levels of CD3. Reciprocal to the decrease of HSA or 6C10/Thy-1, 6C10-(HSA-)3G11+ cells show increased expression of CD45RB/C and Mel-14 reaching levels present on the majority of peripheral T cells, in association with their functional progression. Thus, these data define a thymic CD4+ T-cell differentiation pathway: from HSA+6C10+ to HSA^low6C10-, finally HSA-6C10-, all retaining 3G11 expression, as on peripheral naive T cells. Both HSA^low and HSA-(3G11+) cells migrate into periphery as 6C10-3G11+ cells, establishing the naive T-cell pool, during which HSA expression is finally lost.

Spleen

Most CD4+ T cells in the murine periphery (spleen, lymph nodes, blood) simultaneously express CD45RB/C, Mel-14, and 3G11 [11, 13, 25, 26]. In agreement with reports fractionating cells based on CD45RB or Mel-14, cells which predominantly secrete IL-2 upon stimulation are found in two 3G11+ cell subsets in the periphery (Fr. I and II), fractions shown in figure 1c. The finding of two naive subsets is unique to the mouse system, providing further insight on CD4+ T-cell differentiation. Our subsequent data has suggested that Fr. I represents a population of cells most recently exported from the thymus, since: (1) it is most sensitive to thymic depletion treatments, such as adult thymectomy; (2) it is the first cell subset appearing after depletion of peripheral CD4+ T cells by in vivo anti-T-cell antibody treatment, and (3) it is the only naive cell fraction present in neonatal mice, with Fr. II cells accumulating later in development [21]. Such progression from Fr. I to II appears to be antigen-independent. This was demonstrated by using transgenic mice with specificity to a foreign antigen. As found in normal adult mice, Fr. II comprises the major fraction of peripheral CD4+ T cells of adult TCR α+β+ transgenic mice with a specificity for pigeon cytochrome C/I-E^k in the absence of deliberate antigen administration [B. Fazekas de St. Groth, Stanford University, pers. commun.]. This was also observed with TCR β chain transgenic mouse of the same specificity as shown below. Thus, we suggest that 6C10/Thy-1 (but not HSA) is up-regulated on peripheral CD4+ T cells in an antigen-independent manner.

Besides the difference of 6C10/Thy-1 levels, the expression of several other molecules, such as CD4 or Ly6C, is also increased on Fr. II compared to Fr. I [21]. It is not clear at present whether such altered expression has biological significance relating to cell function in all cases. However, it is interesting to note that there is a functional distinction between the phenotypically resolvable subsets, Fr. I and II. In a Con-A stimulation system, while both resting B cells and non-B cells can serve as accessory cells for Fr. II, resting B cells are unable to provide accessory signals for Fr. I [13]. Furthermore, we found that helper function for the antibody response was more efficient with Fr. II cells and that early memory helper T cells were found in Fr. II (and also Fr. III). In contrast, no such helper function was found with Fr. I [13]. These data suggest the importance of further progression of CD4+ T cells in the periphery, acquiring certain functional competence.

Presence of Cell Subsets in the Thymus and Periphery with Unique Cell Surface Phenotype and Cytokine Profiles

As figure 1 shows, besides the two major 3G11+ subsets, 3G11− subsets are present as minor fractions in both thymus and spleen. Their phenotype and cytokine spectrum distinguishes them from any 3G11+ fractions. Tables 1 and 2 summarize their phenotypic and functional features.

3G11− Cells in the Thymus (Mthy4)
Both 3G11+ and 3G11− cells are found in the most mature 6C10−(HSA−) fraction of thymus (fig. 1b). Unlike the 3G11+ subset, whose phenotype resembles the peripheral Fr. I naive subset, the 6C10−3G11− subset includes CD3low cells (or consists entirely of CD3low cells), is Mel-14−, bears low levels of Thy-1, high levels of CD44, and levels of CD45RB/C that are slightly higher than peripheral cells. Although the average cell size is somewhat larger than other CD4+ T cells in the thymus, they do not express IL-2 receptor, different from recently activated cells. In response to anti-CD3 stimulation, they show rapid activation and secrete high levels of IL-4, IL-5, and IFN-γ in short-term culture and lower levels of IL-2. This subset with such a distinctive cytokine spectrum, termed ThyO, is detectable in the thymus of neonatal mice, increases during development into adulthood, where it comprises a large fraction of the HSA− fraction, a subset already known to secrete diverse cytokines. Resolution of ThyO from the authentic intrathymic differentiation pathway of CD4+ T cells is demonstrated by the age-dependent altera-

Table 1. Cell surface phenotype of CD4$^+$ T-cell subsets

Location	Subset	CD3	CD45RB/C	Mel-14	CD44	Ly6A/E[1]	Ly6C[1]
Thymus	3G11$^+$6C10$^-$	++	+	+	+	ND	ND
	3G11$^-$6C10$^-$ (ThyO)	+	+	–	++	ND	ND
Spleen	3G11$^+$6C10$^-$ (Fr. I)	++	+	+	+	+	+
	3G11$^+$6C10$^+$ (Fr. II)	++	+	+	+	+	++
	3G11$^-$6C10$^+$ (Fr. III)	+	–	–	++	++	–
	3G11$^-$6C10$^-$ (Fr. IV)	+	–	–	++	–	–

[1] Tested in B6 mice (BALB/c-negative).

Table 2. Cytokine spectra after anti-CD3 stimulation

Location	Subset	IL-2	IL-4	IL-5	IFN-γ
Thymus	3G11$^+$6C10$^-$	+	+/–	–	–
	3G11$^-$6C10$^-$ (ThyO)	+/–	++	++	++
Spleen	3G11$^+$6C10$^-$ (Fr. I)	++	–	–	–
	3G11$^+$6C10$^+$ (Fr. II)	++	+/–	–	+/–
	3G11$^-$6C10$^+$ (Fr. III)	+/–	++	++	++
	3G11$^-$6C10$^-$ (Fr. IV)	–	+/–	–	+/–

tion of ThyO repertoire where TCR Vβ8$^+$ cells are overrepresented (40–60%) in adult, but not neonatal mice. While no other CD4$^+$ T cell subset in thymus or spleen shows such bias through ontogeny, the CD4$^-$CD8$^-$ HSA$^-$ TCR αβ$^+$ T-cell fraction in thymus does show a profound Vβ8$^+$ overexpression which is also age-dependent. Thus, it is probable that ThyO is generated by intrathymic stimulation and may represent an intermediate stage of cell maturation in the process of eventual CD4 loss [Hayakawa et al., submitted].

3G11$^-$ Cells in Periphery (Fr. III and IV)
In the case of the peripheral CD4$^+$ T cells, two 3G11$^-$ T-cell subsets (Fr. III and IV) are resolved by 6C10 expression (fig. 1c). The majority of cells in Fr. III and IV share the phenotype CD44^{high}, Mel-14$^-$, CD45RB/C$^-$, Ly6C$^-$,

and CD3[low] [21]. In addition to cell surface immunofluorescence staining data, mRNA data also demonstrated decreased usage of CD45RB and CD45RC exons by both 3G11⁻ cell subsets [P. Rogers et al., submitted]. Cell size in both subsets does not distinguish them from the 3G11⁺ subsets and both are in G0/G1 stage [present authors, unpubl.]. Despite this phenotypic similarity between Fr. III and IV, we found that they are functionally distinct. In response to anti-CD3, Fr. III cells respond with limited proliferation and secrete diverse cytokines (IL-4, IL-5, IFN-γ) with less IL-2. In contrast, Fr. IV cells do not respond to such stimulation, although if any lymphokines are detected, these are IL-4 and IFN-γ [21]. Furthermore, long-term helper memory function was restricted to cells with Fr. III phenotype and Fr. IV cells never exhibited such function [15]. Fr. IV cells are in an anergic state since their nonresponsiveness can be overcome to a degree by PMA+Ca^{++} stimulation or by exogenous IL-2 supplement [present authors, unpubl.].

3G11⁻ Cell Subsets (ThyO, Fr. III and IV) Are Generated by Antigenic Selection

As described above, all 3G11⁻ cells show common phenotypic and functional features. That is, they can be recognized as subsets which show a unique cell surface phenotype different from the bulk of CD4⁺ T cells, such as CD44(Pgp-1)[high], Mel-14⁻ and CD3[low]. They all show little or no IL-2 secretion upon stimulation resulting in limited cell proliferation, but reciprocally show more diverse cytokine profiles. Antigenic selection to generate such cells is suggested for ThyO in the thymus by the increase in frequency of Vβ8⁺ cells with development as described above. However, such a difference in repertoire is not obvious from such analysis of Fr. III and IV in normal mice.

The idea that the Fr. III subset might also be composed of secondary cells was foreshadowed by several observations. Besides 3G11, whose expression is permanently lost following antigenic stimulation [15], the levels of other molecules (CD44, Mel-14, CD45RB/C, Ly6C) expressed on Fr. III (and Fr. IV) resemble the secondary cells eventually generated by activation in vitro [12, 26–29]. Generation of 3G11⁻ Fr. III cells from the 3G11⁺ Fr. II cell stage is suggested by the fact that cells with helper memory function are found initially in both 3G11⁺ and 3G11⁻ subsets (Fr. II and III) eventually becoming restricted exclusively to 3G11⁻ cells (Fr. III) long after antigen priming [13]. During ontogeny, the frequency of the Fr. III subset increases with age [21, 30]. These data suggest that a similar mechanism may have operated to generate Fr. III by natural antigenic selection and so accumulating with ageing. In contrast, the origin of Fr. IV is more speculative. Fr. IV cells are already present in neonatal mice but do not show significant increase with age, unlike Fr. III. Since cell activation of 6C10/3G11$^+$ cells can induce loss of both 6C10 and 3G11 expression in association with the loss of responsiveness, we presume that Fr. IV found in normal mice may represent a subset at a terminal stage of CD4$^+$ differentiation after cell activation. Immediate precursor cells for Fr. IV could be any CD4$^+$ subsets in the periphery or even ThyO in the thymus.

Higher Endogenous β Gene Usage by the 3G11$^-$ Cell Subsets of TCR β Chain Transgenic Mice

To overcome difficulties in investigating distinctions of TCR repertoire in 3G11$^-$ subsets, we have utilized the 2B4βE$_H$ mouse, transgenic for the β chain of a TCR encoding the specificity for a foreign antigen/class II restricted (pigeon cytochrome C/I-E^k) [31]. This approach is based on our assumption that if such 3G11$^-$ cell subsets are selected by antigen (presumably endogenous antigens): (1) the presence of such subsets may be rare in the transgenic mice since they have less chance of expressing an endogenous antigen specificity, and (2) if they are present, they may be enriched for cells using nontransgene (endogenous) β chains combined with a restricted usage of endogenous TCR α chains. As shown in figure 3, our preliminary data supports our prediction. ThyO is present, but its frequency is greatly reduced among CD4$^+$ T cells in the thymus of 2B4βE$_H$ mouse compared to normal mice. In comparison, four subsets are present in the spleen, at levels only slightly decreased compared with nontransgenic animals. 3G11$^-$ cells in these mice show typical phenotype as found in normal mice such as CD44^{high} and CD45RB^{low} (fig. 3). All 3G11$^-$ subsets show the characteristic response with respect to cytokine secretion as found in normal mice. As expected, the transgene (Vβ3) is expressed on the majority of 3G11$^+$ cell subsets in the thymus and spleen, presumably combined with random usage of TCR α chains as unselected naive cells. The majority of ThyO cells are also Vβ3 transgene$^+$; however, some Vβ8$^+$ cells are uniquely found in the ThyO but not in any other cell subsets (data not shown). Distinctively, as shown in figure 4, a large proportion (50%) of Fr. III cells (in both spleen and lymph nodes) do not express the transgene. The contribution of such cells lacking transgene is still significant in Fr. IV, although to a lesser extent than Fr. III. Although our study requires further experiments, in particular determination of TCR α

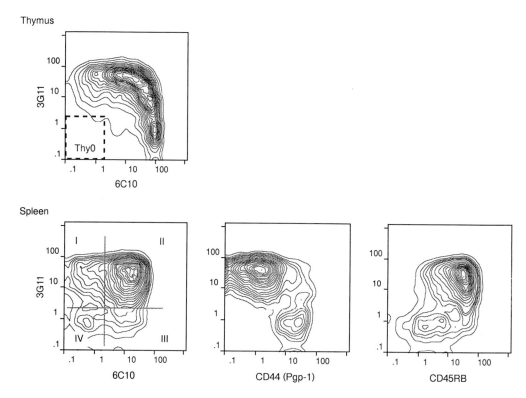

Fig. 3. CD4⁺ T-cell subsets in the 2B4βE$_H$ transgenic mouse. The original founder of the 2B4βE$_H$ transgenic mouse was backcrossed onto B10.BR mice for several generations, and the mouse used here is H-2^k (and Mls2,3^b, free of deletional effect). CD4⁺CD8⁻ cells comprise 17% of thymocytes and 13% of spleen cells in this mouse. ThyO is only 3% of CD4⁺ T cells in the thymus. Subsets in the spleen are demarcated. Fr. III and IV subsets are 6 and 8% of splenic CD4⁺ T cells, respectively. 3G11⁻ subsets are mostly CD44(Pgp-1)^{high}, and lower for CD45RB in comparison with 3G11⁺ cells.

chain usage, these data strongly suggest that the mechanisms to generate 3G11⁻ subsets are all principally based on antigenic selection.

Summary: CD4⁺ Subsets in Relation to the Cell Differentiation and Maturation Pathways

As shown in figure 2, these phenotypically identifiable subsets represent cell progression stages occurring in the thymus or in the periphery, either

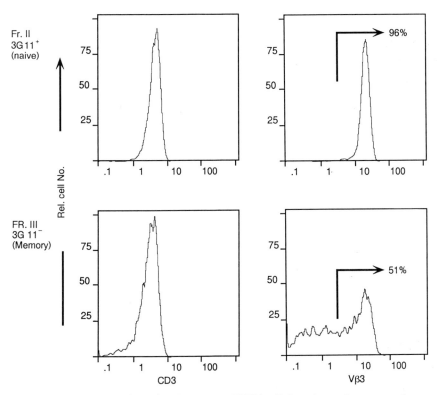

Fig. 4. Frequency of CD3⁺ and transgene (Vβ3)⁺ cells in naive and memory subsets. Spleen cells were simultaneously stained with antibodies to CD4, 6C10, 3G11, and CD3 or Vβ3 (KJ25). CD3 or Vβ3 histograms of Fr. II and III subsets are shown. As previously described, CD3 levels on 3G11⁻ cells are lower (but all positive) than on 3G11⁺ cells.

independent of antigen (retaining 3G11 expression) or antigen dependently (losing 3G11). Although Fr. III is likely to be derived from Fr. II cells, a direct precursor for Fr. IV cells remains unclear. Since Fr. IV cells are already present in neonatal mice prior to the appearance of Fr. III and II in the periphery, Fr. IV cells could derive either from Fr. I in the periphery or from ThyO exported from the thymus as a terminal stage. We expect that antigens involved in generating the three types of 3G11⁻ secondary subsets in the thymus or in the periphery may differ, as suggested by data on transgenic mice. All these secondary 3G11⁻ subsets are characterized by secretion of cytokines other than IL-2.

Epilogue

Our investigations described here have been to answer the question of whether CD4+ subsets characterized by distinct cytokine spectra are generated by antigen-independent differentiation process or not. Since our data show such subsets are all secondary, it provides further evidence to support the idea of lymphocyte maturation. That is, T-cell maturation characterized by cytokine switching and alteration of cell surface phenotype (in relation with T-cell antigen recognition) in analogy with B cells where maturation is recognized by the immunoglobulin (Ig) isotype switching and Ig affinity increase by somatic mutation. However, the question of generation of CD4+ T cells committed to either Th1 or Th2 (or other) type cells remains unanswered in this scheme and so further study is required. We assume that such commitment occurs after antigenic exposure and that regulatory mechanisms play an important role as proposed by others [6, 25, 32, 33]. A number of interesting experimental approaches are in progress to resolve this question and we hope to achieve a fuller understanding in the future.

Acknowledgments

We wish to thank Dr. M. Davis (Stanford University) for providing 2B4βE$_H$ transgenic mice, and Drs. R.L. Coffman (DNAX) and K. Bottomly (Yale University) for providing hybridomas or antibody to carry out CD4+ T-cell subset study described here. This work was supported by grants from the National Institutes of Health (CA-06927,RR-05539), the Pew Charitable Trust (86-5043HE), the Pew Charitable Trust Five Year Award (83-1067HE), and by an appropriation from the Commonwealth of Pennsylvania.

References

1 Cantor H, Shen FW, Boyse EA: Separation of helper from suppressor T cells expressing Ly components. II. Activation by antigen: after immunization, antigen-specific suppressor and helper activities are mediated by distinct T-cell subclasses. J Exp Med 1976;143:1391–1401.
2 Dialynas DP, Wilde DB, Marrack P, Pierres A, Wall KA, Havran W, Otten G, Loken MR, Pierres M, Kappler J, Fitch FW: Characterization of the murine antigenic determinant, designated L3T4a, recognized by monoclonal antibody GK1.5: Expression of L3T4a by functional T cell clones appears to correlate primarily with class II MHC antigen reactivity. Immunol Rev 1983;74:29–56.
3 Morimoto C, Letvin NL, Distaso JA, Aldrich WR, Schlossman SF: The isolation and characterization of the human suppressor inducer T cell subset. J Immunol 1985;134:1508–1515.

4   Morimoto C, Letvin NL, Boyd AW, Hagan M, Brown HM, Kornacki MM, Schloss-
     man SF: The isolation and characterization of the human helper inducer T cell
     subset. J Immunol 1985;134:3762–3769.
5   Mosmann TR, Cherwinski H, Bond MW, Giedlin MA, Coffman RL: Two types of
     murine helper T cell clone. I. Definition according to profiles of lymphokine
     activities and secreted proteins. J Immunol 1986;136:2348–2357.
6   Mosmann TR, Coffman RL: TH1 and TH2 cells: Different patterns of lymphokine
     secretion lead to different functional properties. Annu Rev Immunol 1989;7:145–
     173.
7   Street NE, Schumacher JH, Fong AT, Bass H, Fiorentino DF, Leverah JA, Mosmann
     TR: Heterogeneity of mouse helper T cells. J Immunol 1990;144:1629–1639.
8   Janeway CA Jr, Carding S, Jones B, Murray J, Portiles P, Rasmussen R, Rojo J,
     Saizawa K, West J, Bottomly K: CD4⁺ T cells: Specificity and function. Immunol
     Rev 1988;101:39–79.
9   Sanders ME, Makgoba MW, Shaw S: Human naive and memory T cells: reinterpreta-
     tion of helper-inducer and suppressor-inducer subsets. Immunol Today 1988;9:195–
     199.
10   Arthur RP, Mason D: T cells that help B cell responses to soluble antigen are
     distinguishable from those producing interleukin-2 on mitogenic or allogeneic
     stimulation. J Exp Med 1987;163:774–786.
11   Bottomly K, Luqman M, Greenbaum L, Carding S, West J, Pasqualini T, Murphy
     DB: A monoclonal antibody to murine CD45R distinguishes CD4 T cell populations
     that produce different cytokines. Eur J Immunol 1989;19:617–623.
12   Budd RC, Cerottini J-C, Horvath C, Bron C, Pedrazzini T, Howe RC, MacDonald
     HR: Distinction of virgin and memory T lymphocytes. Stable acquisition of the
     Pgp-1 glycoprotein concomitant with antigenic stimulation. J Immunol 1987;138:
     3120–3129.
13   Hayakawa K, Hardy RR: Murine CD4⁺ T cell subsets defined. J Exp Med 1988;
     168:1825–1838.
14   Powers GD, Abbas AK, Miller RA: Frequencies of IL-2- and IL-4-secreting T cells in
     naive and antigen-stimulated lymphocyte populations. J Immunol 1988;140:3352–
     3357.
15   Hayakawa K, Hardy RR: Phenotypic and functional alteration of CD4⁺ T cells after
     antigen stimulation: resolution of two populations of memory T cells that both
     secrete IL-4. J Exp Med 1989;169:2245–2250.
16   Serra HM, Krowka JF, Ledbetter JA, Pilarski LM: Loss of CD45R (Lp220) repre-
     sents a post-thymic T cell differentiation event. J Immunol 1988;140:1435–1441.
17   Akbar AN, Terry L, Timms A, Beverley PCL, Janossy G: Loss of CD45R and gain of
     UCHL1 reactivity is a feature of primed T cells. J Immunol 1988;140:2171–2178.
18   Cerottini JC, MacDonald HR: The cellular basis of T-cell memory. Annu Rev
     Immunol 1989;7:77–89.
19   Firestein GS, Roeder WD, Laxer JA, Townsend KS, Weaver CT, Hom JT, Linton J,
     Torbett BE, Glasebrook AL: A new murine CD4⁺ T cell subset with an unrestricted
     cytokine profile. J Immunol 1989;143:518–525.
20   Hayakawa K, Carmack CE, Hyman R, Hardy RR: Natural autoantibodies to
     thymocytes: origin, VH genes, fine specificities, and the role of Thy-1 glycoprotein.
     J Exp Med 1990;172:869–878.
21   Hayakawa K, Hardy RR: Murine CD4⁺ T cell subsets. Immunol Rev 1991;123:145–168.

22 Crispe N, Bevan MJ: Expression and functional significance of the J11d marker on mouse thymocytes. J Immunol 1987;138:2013–2018.

23 Wilson A, Day LM, Scollay R, Shortman K: Subpopulations of mature murine thymocytes: Properties of CD4⁻CD8⁺ and CD4⁺CD8⁻ thymocytes lacking the heat-stable antigen. Cell Immunol 1988;117:312–326.

24 Bruce J, Symington FW, McKearn TJ, Sprent J: A monoclonal antibody discriminating between subsets of T and B cells. J Immunol 1981;127:2496–2501.

25 Swain SL, Bradley LM, Croft M, Tonkonogy S, Atkins G, Weinberg AD, Duncan DD, Hedrick SM, Dutton RW, Huston G: Helper T cell subsets: phenotype, function and the role of lymphokines in regulating their development. Immunol Rev 1991;123:115–144.

26 Birkeland ML, Johnson P, Trowbridge IS, Pure E: Changes in CD45 isoform expression accompany antigen-induced murine T-cell activation. Proc Natl Acad Sci USA 1989;86:6734–6738.

27 Hamann A, Jablonski-Westrich D, Scholz K-U, Duijvestijn A, Butcher EC, Thiele H-G: Regulation of lymphocyte homing. I. Alterations in homing receptor expression and organ-specific high endothelial venule binding of lymphocytes upon activation. J Immunol 1988;140:737–743.

28 Jung TM, Gallatin WM, Weissman IL, Dailey MO: Down-regulation of homing receptors after T cell activation. J Immunol 1988;141:4110–4117.

29 Rock KL, Reiser H, Bamezai A, McGrew J, Benacerraf B: The Ly-6 locus: A multigene family encoding phosphatidylinositol-anchored membrane proteins concerned with T-cell activation. Immunol Rev 1989;111:195–224.

30 Ernst DN, Hobbs MV, Torbett BE, Glasebrook AL, Rehse MA, Bottomly K, Hayakawa K, Hardy RR, Weigle WO: Differences in the expression profiles of CD45RB, Pgp-1, and 3G11 membrane antigens and in the patterns of lymphokine secretion by splenic CD4⁺ T cells from young and aged mice. J Immunol 1990;145:1295–1302.

31 Berg LJ, Pullen AM, Fazekas de St Groth B, Mathis D, Benoist C, Davis MM: Antigen/MHC-specific T cells are preferentially exported from the thymus in the presence of their NHC ligand. Cell 1989;58:1035–1046.

32 Gajewski TF, Schell SR, Nau G, Fitch FW: Regulation of T-cell activation: Differences among T-cell subsets. Immunol Rev 1989;111:79–110.

33 Le Gros G, Ben-Sasson SZ, Seder R, Finkelman FD, Paul WE: Generation of interleukin-4 (IL-4)-producing cells in vivo and in vitro: IL-2 and IL-4 are required for in vitro generation of IL-4 producing cells. J Exp Med 1990;172:921–929.

Kyoko Hayakawa, MD, PhD, Institute for Cancer Research,
Fox Chase Cancer Center, 7701 Burholme Ave., Philadelphia, PA 19111 (USA)

Coffman RL (ed): Regulation and Functional Significance of T-Cell Subsets.
Chem Immunol. Basel, Karger, 1992, vol 54, pp 117–135

Induction, Regulation and Function of T-Cell Subsets in Leishmaniasis

F.Y. Liew

Department of Immunology, University of Glasgow, Western Infirmary, Glasgow, UK

Heterogeneity of CD4 T Cells

By 1970 it was established that lymphocytes could be divided into B and T cells, and that T cells help B cells to produce antibodies. By the mid-1970s it was clear that the helper T cells which are Lyt-1^+2^- were distinct from the killer cells which are Lyt-1^-2^+. However, even before the identification of T and B cells, there was evidence that the immune response could be divided into humoral and cell-mediated immunity and that these two responses did not always appear in parallel. It was demonstrated that animals pretreated with antigen in saline failed to develop delayed-type hypersensitivity (DTH), a typical cell-mediated immunity, to a subsequent challenge of the same antigen in adjuvant, but produced normal or variable levels of antibody responses [1–9]. This phenomenon has been referred to as 'immune deviation' [2], 'preimmunization tolerance' [6] or 'split tolerance' [7]. In contrast, normal development of DTH was maintained in animals made antibody tolerant to tuberculin [10]. Parish [11] first showed that there can be an inverse relationship between humoral and cell-mediated immunity. This was followed by the demonstration [12] that this phenomenon extended to high and low zone antibody tolerance (fig. 1). However, it was not until 1974 that it was shown that the T cells helping antibody synthesis and those mediating DTH were distinct [13, 14]. The heterogeneity of Lyt-1^+ (CD4) cells was further sustained by the demonstration that antigen-specific helper T cells and nonspecific helper T cells could be segregated in vitro by limiting dilution [15], and that two populations of helper T cells were needed to provide optimal help in antibody responses [16]. This was followed by a period during which much of the attention of cellular immunologists was

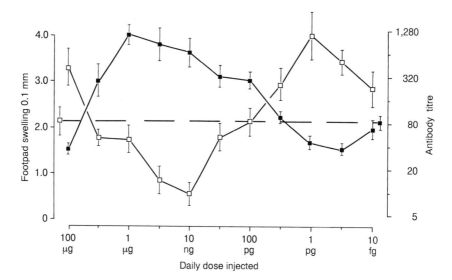

Fig. 1. Inverse relationship between humoral and cell-mediated immunity. Groups of 7 rats (strain W, Wistar) were injected intraperitoneally daily for 27 days with doses of a CNBr digest of flagellin (*Salmonella adelaide*, strain SW 1338) varying in 10-fold dilution steps from 100 μg to 10 fg (fg = femtogram). All animals, together with unimmunized controls, were challenged on day 28 with 100 μg of flagellin in 100 μl saline in the right hind footpad. The antibody response (■) represents the mean of the 7-, 14-, 21- and 28-day post-challenge serum antibody titres (reciprocal of dilutions) determined by hae-magglutination using sheep erythrocytes sensitized with polymerized flagellin. DTH (□) was elicited in the left hind footpad with 0.5 μg flagellin in 100 μl saline 28 days after the flagellin challenge. DTH, expressed as 24 h footpad swelling, was determined by the difference in the footpad thickness between left and right hind footpad. The broken line represents the antibody and DTH responses of control rats which were injected only with 100 μg of flagellin in saline on day 28. Vertical bars represent standard errors of the means [from 12].

directed at the elucidation of suppressor T cells, a cell type which still awaits clarification.

The next stage of the development in the study of CD4 T-cell heterogeneity came from the investigation of the immune response to the protozoan parasite *Leishmania*. BALB/c are highly susceptible to *Leishmania major* infection. They develop uniformly fatal visceral disease even with a minimal infective dose. However, when these mice were exposed to a sublethal dose of γ-irradiation shortly before infection, they were able to control the disease and eventually achieve complete healing. This acquired resistance can be reversed by transferring into the irradiated recipients T cells from normal donors or, even more effectively, T cells from BALB/c mice with progressive disease [17]. Both the protective T cells [17, 18] and the disease-exacerbating T cells are CD4+ [19]. The disease-promoting cells were cloned [20] and found to be extremely potent; as few as 10^6 cells can reverse the disease-resisting capability of γ-irradiated BALB/c mice. The host-protective T cells mediate DTH whereas the disease-exacerbating T cells suppress DTH. These results therefore demonstrate clearly that they are two functionally distinct populations of CD4 T cells. However, the concept of Th1 and Th2 cells was not generally accepted until the discovery that CD4 cells can be separated according to the pattern of lymphokines they produce [21]. As discussed in detail elsewhere in this volume, the principal contrasting characteristics of the two subsets of CD4 T cells are: Th1 cells produce IL-2, IFN-γ and mediate DTH, whereas Th2 cells secrete IL-4, IL-10 and are essential for the synthesis of IgE. This article will review the current understanding of the induction, regulation and effector mechanisms of CD4 T-cell subsets in leishmaniasis.

Leishmaniasis

Leishmaniasis is caused by species of the intracellular protozoan parasite belonging to the genus *Leishmania*. There are three main categories of leishmaniasis.

Cutaneous leishmaniasis caused by *L. major* (Oriental sore) is found mainly in the Middle East and Africa, while those caused by *L. mexicana, L. amazonensis, L. panamensis* (chiclero's ulcer or bay sore) occur in South America. Most cutaneous leishmaniases produce a skin ulcer which heals spontaneously leaving an unsightly scar. Control of the disease is attributed to acquired cell-mediated immunity which is detectable as DTH in vivo and by blast transformation and inhibition of macrophage migration in vitro. Serum antibody titres are low in general and directly related to severity of the disease. Another form of cutaneous leishmaniasis, diffuse cutaneous leishmaniasis, occurs both in the Old and New Worlds. This causes widespread thickening of the skin with lesions that resemble those of lepromatous leprosy which fail to heal spontaneously. This form is characterized by the persistence of low numbers of parasites within the macrophages, frequently in the presence of strong DTH, suggesting an immunological defect in the host.

Mucocutaneous leishmaniasis (espundia) caused by *L. braziliensis* is due to metastasis of organisms to mucosal sites from a primary cutaneous lesion established much earlier. Metastatic spread to the oronasal and pharyngeal mucosa may occur, causing highly disfiguring leprosy-like tissue destruction and swelling. The incidence of this spread and its pathogenesis is unclear, since both DTH and serum antibody are present.

Visceral leishmaniasis (kala azar) caused by *L. donovani* is accompanied by weight loss, anaemia, skin darkening (black disease) and hepatospleno-megaly. The disease is usually fatal if untreated. Kala azar is characterized by high titres of specific antibodies and polyclonal B-cell activation but profound impairment of DTH and blastogenic response. Failure of cell-mediated immunity in kala azar is related to the inability to eliminate the parasite, since individuals cured by chemotherapy subsequently express both strong DTH and resistance to reinfection.

With the exception of Oceania, leishmaniases occur in most parts of the world. Overall, about a third of the world's population (some 1.6 billion people) is at risk of infection and disease. An incidence of 400,000 new cases per year has been reported and the world-wide prevalence of the disease is believed to be about 12 million cases [22]. Treatment of leishmaniasis is based on antimony compounds, notably the pentavalent antimonials sodium stibogluconate and N-methylglucamine antimonate, but they have to be given in daily intramuscular doses for several weeks with unpleasant side-effects and are not effective against cutaneous forms. Progress in this field has not been impressive.

Vaccination remains perhaps the best answer to the control of leishma-niasis. However, so far, the only immunization strategy against leishmaniasis used in man with any success has been restricted to the cutaneous disease. It is based on convalescent immunity following controlled infection in an aestheti-cally acceptable site with viable *L. major* [23]. This process called leishmaniza-tion is effective but flawed with complications and hence will not be acceptable for general use. However, it demonstrates that vaccination against leishmania-sis is feasible in principle. Recent advances in the knowledge of the immunol-ogy of leishmaniasis will greatly enhance this possibility.

Host-Protective Immune Response

In experimental murine models, significantly higher levels of specific and polyclonal antibodies are present in mice with progressive disease than the mice recovered from the infection [24]. Antileishmanial antibodies have been shown in vitro to lyse promastigotes in the presence of complement [25] and to promote phagocytosis [26]. However, there is no corresponding in vivo role for antibody in determining the outcome of leishmanial infection. Inbred mouse strains with widely different degrees of susceptibility produce antibodies similar in titres and isotypes following infection with *L. major* [27]. Furthermore, the disease progression in mice infected with *L. major* was not modified with the passive transfer of large amounts of specific antibody (9 ml/mouse over 5 weeks) from mice protectively immunized with killed promastigotes [28]. There is evidence that individuals infected with *L. braziliensis* produce significantly higher levels of total IgE compared to normal controls [29]. However, the specificity and the relevance of the antibody are unknown. Higher levels of specific IgE are detected in mice with progressive disease than in mice resistant to *L. major* infection [30]. The enhanced IgE response is consistent with the presence of Th2 cells in these mice but its contribution to the disease outcome has not been fully evaluated.

Treatment of mice from birth with anti-IgM antibody, a process known to abrogate the antibody response, can profoundly affect the outcome of *L. major* infection [31, 32]. This effect, however, is likely to be due to the depletion of the antigen-presenting function of B cells rather than the reduction of antibody production, since the effect of IgM treatment can be reversed by the transfer of T cells but not by specific antibody.

In contrast to antibody, the case for cell-mediated immunity playing a pivotal role in vivo is based on a range of impressive clinical and experimental evidence. Resistant mice rendered relatively T-cell deficient by thymectomy followed by irradiation and reconstitution with syngeneic bone marrow cells are less able to control *L. major* infection [33], and athymic mutants of the highly resistant mice are totally unable to control *L. major* infection with uniformly fatal outcome. Normal resistance can be fully restored by reconstitution with normal syngeneic T cells [34]. Acquired immunity against *L. major* [35] and *L. donovani* [36] as a result of recovery from infection or prophylactic immunization can also be transferred by T cells but not by B cells [37]. Treatment of C3H mice from birth with anti-IgM antibody rendered them defective in antibody synthesis and also susceptible to leishmanial infection. However, transfer of T cells alone from normal syngeneic donors can completely reverse the disease progression in these mice without any restoration of antibody formation [32]. Together, these results provide a convincing argument for a central role of cell-mediated immunity in acquired resistance to leishmaniasis.

There is also persuasive evidence that CD4 T cells are primarily involved in conferring protective immunity against leishmaniasis. Studies from several laboratories demonstrated that CD4 T cells from mice recovered from infection [17, 18, 38] or following prophylactic immunization [37] are able to adoptively transfer protection in otherwise susceptible recipients. Macrophage-activating factors such as IFN-γ produced by specifically sensitized T cells are deemed to be essential for the activation of infected macrophages to eliminate intracellular amastigotes [39]. The host-protective T cells are therefore akin to the Th1 cells [52, 53].

Recent evidence suggests that CD8 T cells may also be protective against *L. major* infection [40]. Resistant CBA mice became less able to heal from *L. major* infection after repeated CD8 antibody treatment in vivo. However, the effect of CD8 antibody treatment was far less impressive than that of CD4 antibody. The extreme susceptibility of athymic nude BALB/c mice to *L. donovani* infection could only be reversed by a combination of CD4 and CD8 T cells [41]. Either population alone failed to reconstitute the athymic mice to the resistant status of their euthymic littermates. However, antigen-specific CD8 cytotoxic T cells have not been demonstrated in leishmaniasis. There is now clear evidence that CD8 cells mediate their protective effect in leishmaniasis by producing IFN-γ [42], although CD8 T cells produce, on an average, 10 times less IFN-γ than CD4 T cells.

Suppression of Host-Protective Immunity

Clinically, patients with visceral leishmaniasis fail to develop a leishmanial-specific skin reaction or proliferative T-cell response [43, 44]. These reactions are fully restored following successful chemotherapy [45, 46]. In experimental models, the reduced T-cell response to phytohaemagglutinin during *L. donovani* infection in BALB/c mice compared with that of normal controls is associated with impaired IL-2 production [47]. The depressed mitogen-induced cellular proliferation and IL-2 production in mice susceptible to *L. major* infection is attributable to the presence of a population of macrophage-like adherent suppressor cells [48, 49]. A similar nonspecific suppression of IL-2 production in clinical visceral leishmaniasis has also been demonstrated [50, 51]. The suppression of IL-2 production appears to be mediated by macrophages and is pronounced only at the terminal stages of infection [48, 49]. In contrast, the antigen-specific suppression of DTH occurs earlier and is mediated by CD4 T cells. The identity of these CD4 suppressor T cells has been controversial because their suppressive activity was defined in operational terms. However, recent studies have clearly indicated that they are akin to Th2 cells [52, 53]. T cells from the lymph node of susceptible mice produce predominantly IL-4, and neutralization of IL-4 in vivo allows otherwise susceptible BALB/c mice to heal [54]. T-cell lines which have the Th2 cytokine-producing pattern adoptively mediate disease exacerbation [55]. Lymph node cells from mice with progressive disease preferentially express message for IL-10 [56], a lymphokine characteristic of Th2 cells.

Thus, there appears to be a dichotomy in the immune response to cutaneous leishmaniasis: Th1 cells are host-protective whereas Th2 cells are disease-promoting [52, 53]. However, there are a number of discrepancies. Firstly, it has been reported that some T-cell lines which produce IFN-γ are also disease-exacerbative when adoptively transferred to naive recipients [57, 58]. Secondly, T cells from mice prophylactically immunized with killed parasites, though secreting IFN-γ and host-protective, failed to mediate DTH [37]. In fact they suppress DTH [59]. T cells from mice immunized subcutaneously with killed parasites produce IL-3 and IL-4 and are disease-promoting, yet they mediate the Jones-Mote (DTH) reaction [60]. Thirdly, Th1 and Th2 populations are not clearly demonstrable in murine *L. dono-vani* infection [61]. These results suggest that the heterogeneity of CD4 T cells may extend beyond the current Th1 and Th2 classification, with some yet undefined cytokines playing influential roles in the regulation of immune responses. However, since the function and characteristics of these cells are speculative, I shall confine my discussion to the context of Th1 and Th2 cells whose occurrence in humans is now being more firmly established [62, 63].

Functional Relationship of Th1 and Th2 Cells

It was demonstrated earlier that the culture supernatants of lymphoid cells from BALB/c mice recovered from *L. major* infection following sub-lethal irradiation can activate macrophages to kill intracellular amastigotes [60]. In contrast, the lymphoid cells from BALB/c mice with progressive disease, when stimulated in vitro, produce factors which can inhibit the activation of macrophages [64]. The active ingredient of macrophage-activating factor is predominantly IFN-γ, whereas the inhibiting factors are IL-3 and IL-4 [64]. The whole system can be reproduced with recombinant IFN-γ on the one hand, and IL-3 and IL-4 on the other. Similar results were also

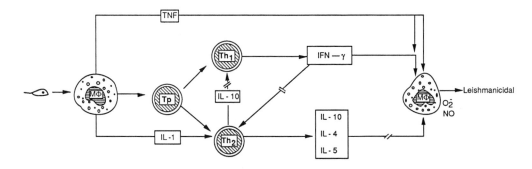

Fig. 2. Schematic representation of the induction and regulation of Th1 and Th2 cells in murine cutaneous leishmaniasis. Mφ = Macrophages, Tp = precursor T cells, NO = nitric oxide, O_2^- = superoxide. Macrophages infected with *Leishmania* can activate Tp to differentiate to Th1 or Th2, they can also produce IL-1 which potentiate the expansion of Th1, and TNF which can synergize with IFN-γ or act directly to activate Mφ. Interrupted arrows represent blockings.

obtained with human monocytes infected with *L. donovani* [65]. Thus it appears that the two subsets of CD4 T cells can modulate the outcome of the disease by influencing the ability of macrophages to kill the intracellular parasites. However, it should be noted that the inhibition of macrophage activation by IL-4 was effective only when the macrophages were pretreated with IL-4 before the stimulation of the cells with IFN-γ. Addition of IL-4 to the cultures after IFN-γ frequently led to enhanced activation [66].

Murine IL-10 is produced by Th2 cells and inhibits the synthesis of cytokines, particularly IFN-γ, by Th1 cells [67]. This inhibitory activity of IL-10 was found to be indirect and required the presence of antigen-presenting cells [68]. Recently, it was demonstrated that IL-10 strongly reduces antigen-specific human T-cell proliferation by diminishing the antigen-presenting capacity of monocytes via down-regulation of class II MHC expression [69]. Thus, this represents another pathway through which Th2 cells regulate the function of Th1 cells [70].

Conversely, it has been reported that IFN-γ can inhibit the proliferation of Th2 cells [71]. The interaction of Th1 and Th2 cells during leishmanial infection is schematically represented in figure 2. This scheme also includes the effect of TNF-α and IL-1β on the activation of macrophages and Th2 cells respectively.

It is now apparent that TNF-α plays an influential role in the control of cutaneous infection in the murine model [72, 73]. It can act directly or synergize with IFN-γ in activating macrophages [74, 75]. Recent evidence suggests that macrophages presenting antigens at high densities and producing IL-1 preferentially activate Th2 cells [76]. This is consistent with the finding that when macrophages are infected with L. major, they are fully competent to induce a T-cell response and to produce significant amounts of IL-1 [77]. This is in contrast to the situation with L. donovani, where the infection of murine macrophages has been shown to reduce the secretion of IL-1 [78]. These contrasting findings reflect the distinct pathology and immune responses caused by the two strains of Leishmania.

Preferential Induction of Th1 and Th2 Cells

The mechanism leading to the preferential induction of Th1 and Th2 cells is at present unclear and is a subject of considerable interest. There are principally three major factors: antigen presentation, distinct antigen recognition repertoire or distinct requirements for differentiation signals.

Earlier studies demonstrated that mice immunized intravenously (i.v.) with killed whole L. major promastigotes or their soluble antigen extract develop substantial resistance to a challenge infection [79]. The protection is mediated by the equivalent of Th1 cells. In contrast, the same antigen preparations injected via the subcutaneous (s.c.) route are not only ineffective but induce a Th2-equivalent cell population which can inhibit the induction of effective protection by the i.v. route of immunization, or the transfer of immunity by protective T cells [80]. These observations underline the crucial role of antigen presentation in the induction of CD4 T-cell subsets. The nature of the antigen presentation and the cell type involved leading to the preferential induction of Th1 and Th2 cells is at present unknown. It may be that this question can only be resolved by detailed analysis of molecularly defined epitopes and their association with class II determinants on monoclonal antigen-presenting cells.

Recently, an octamer of a ten residue peptide corresponding to the tandemly repeating region of a cytosolic protein of L. major was synthesized and tested for its immunogenicity [81]. This peptide (p183), which is H-2^d-restricted, failed to induce protective immunity but can exacerbate disease development in both susceptible and resistant strains of mice. The activated T cells are predominantly CD4 and secrete IL-4, but little or no IL-2 or IFN-γ, when stimulated with soluble parasite extract in vitro. The i.v. route of immunization with p183 is, however, without any detectable effect. These results therefore suggest that p183 preferentially induces Th2 cells leading to disease exacerbation.

Peptides corresponding to the major surface glycoprotein, gp63, have also been synthesized and tested for immunogenicity [82]. A number of these peptides were found to be highly immunogenic in several strains of mice. One of them, p147, is of particular interest. It contains the zinc-binding domain crucial for the enzymatic activity of gp63 [83, 84], a metalloprotease thought to contribute to the survival of *Leishmania* parasites in the hostile environment of macrophage phagolysosomes [85]. Peptide 147 induced predominantly CD4 T cells which secrete IL-2 and IFN-γ, but no detectable IL-4, when stimulated with whole parasite in vitro, suggesting that it may be able to preferentially induce Th1 cells. CBA mice immunized i.v. with p147 developed significant resistance against *L. major* infection compared to controls. However, such immunity can only be achieved by the i.v. route of immunization, the s.c. route was without any detectable effect [82].

These findings therefore demonstrate that both the antigenic determinant and the mode of its presentation play important roles in the induction of CD4 T-cell subsets. There is also clear evidence that certain cytokines can influence the induction of T-cell subsets. For example, when TNF-α was incorporated in the inoculum containing p183 and injected s.c. into BALB/c mice, the mice developed significant resistance to *L. major* infection, instead of disease exacerbation as described earlier [86]. These results suggest that TNF-α can influence the antigen presentation in, perhaps, switching from Th2-inducing mode to that of Th1 induction.

Effector Mechanism of Th1 and Th2 Cells

When normal macrophages are infected with a virulent strain of *Leishmania*, the parasites survive and subsequently replicate in the macrophages until the cells are lysed. In contrast, when the macrophages are activated in vitro with IFN-γ in the presence of low levels of LPS (10 ng/ml), the parasites are killed. It was assumed for a long time that reactive oxygen intermediates such as superoxide and hydrogen peroxide were the major killing mechanism. However, recent studies from several laboratories show that nitric oxide (NO) is in fact the principal effector mechanism in this and other related systems.

NO is involved in a variety of biological activities, including endothelium-related vascular relaxation, platelet aggregation, neurotransmission

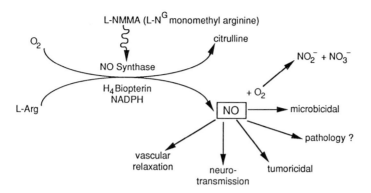

Fig. 3. The arginine:NO pathway.

and macrophage killing of tumour cells (fig. 3) [for review, see 87]. NO is derived from *L*-arginine and molecular oxygen with *L*-citrulline being the co-product. The reaction is catalyzed by the enzyme NO synthase with NADPH, FAD and tetrahydrobiopterin as co-factors. The reaction can be specifically inhibited by *L*-arginine analogues such as *L*-N^G-monomethylarginine (*L*-NMMA). NO is very unstable, having a half-life of 3–15 s. Thus it is usually measured as NO$_2^-$ in the culture supernatants.

Macrophages incubated with medium alone produce little or no detectable NO. However, substantial amounts of NO are generated when the cells are cultured with IFN-γ, TNF-α, or better still, with a combination of the two cytokines, because they synergize with each other in this system. The NO production can be quantitatively inhibited by *L*-NMMA in a dose-dependent manner, but not by the *D*-enantiomer, *D*-NMMA. The production of NO by activated macrophages is directly correlated with leishmanicidal function [88–90]. Macrophages cultured with IFN-γ and TNF-α are highly leishmanicidal. This is not affected by *D*-NMMA, but can be completely inhibited by *L*-NMMA. These results therefore suggest that NO is necessary and may be sufficient to account for the killing of the intracellular pathogens. Furthermore, it has now been shown that NO can kill *L. major* promastigotes in vitro in a cell-free system [89].

The role of NO can also be demonstrated in vivo [89]. BALB/c mice were infected in the footpad with 10^6 *L. major* promastigotes and, when the lesions were developing, *L*-NMMA was injected into the lesions. Mice injected with *L*-NMMA developed significantly larger lesions and 4 orders of magnitude higher parasite load compared with controls injected with *D*-NMMA or PBS. The resistance to *L. major* infection appears to correlate with the ability of macrophages to produce NO. A number of inbred mouse strains which differ substantially in their susceptibility to *L. major* were investigated. Macrophages from the resistant strains expressed significantly higher level of NO synthase and produced larger amounts of NO compared to the susceptible strains when activated with IFN-γ [91]. The genetic control of resistance to pathogens is clearly complex and ill-defined. The induction of NO synthase could be an important contributing factor or the manifestation of a cascade of events.

The next important question is how the activation of macrophages is regulated. There is now clear evidence that IL-4 can inhibit the expression of NO synthase and NO production by IFN-γ-activated macrophages, provided the cells were pretreated with IL-4. Addition of IL-4 after IFN-γ frequently led to enhanced NO synthesis and higher level of leishmanicidal activity [92]. Apart from the cytokines discussed above, a number of other cytokines have now also been found to influence the expression of NO synthase and the production of NO by macrophages. Macrophage migration-inhibiting factor (MIF) has now been cloned and expressed [93]. Recombinant MIF is found to be a potent inducer of macrophage leishmanicidal activity [94], and also induces the expression of NO synthase and the production of NO in these cells [unpubl. data]. In contrast, TGF-β [95] and IL-10 [unpubl. data] have now been shown to be inhibitors of NO synthesis in activated macrophages.

Thus, the following conclusion may be drawn (fig. 4): IFN-γ, TNF-α, MIF and possibly other cytokines, occupying their respective receptors on the macrophages send a series of signals which, together with those of co-stimulators, LPS and perhaps PMA, lead to the induction of NO synthase. Other cytokines such as IL-3, IL-4, TGF-β and IL-10 also send a series of signals which act in the opposite direction, inhibiting the expression of NO synthase. This phenomenon is consistent with the sophisticated regulatory mechanism characteristic of important biological systems. The check-and-balance of two opposing pathways would ensure that this nonspecific cytolytic effector is not overplayed leading to possible pathology [96]. The high degree of degeneracy would ensure that the total breakdown of either system is extremely remote.

The constitutive NO synthase of the brain has now been cloned [97], and the complete amino acid sequence of the inducible NO synthase of the macrophage is now available [D. Bredt and J. Cunningham, pers. commun.]. The unravelling of the signal cascade and the control expression of the

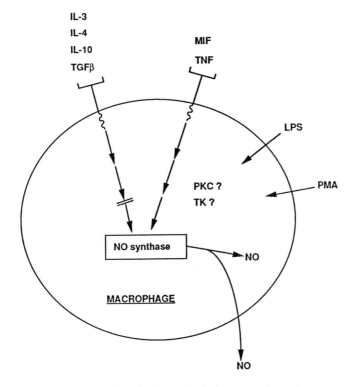

Fig. 4. The regulation of NO synthesis in macrophages by two sets of opposing cytokines.

enzymes and their co-factors will be investigated next. This is a fascinating area with many important questions to be answered and will thus be an extremely rewarding area of research in the next few years. The implication of this information for the control of infectious and autoimmune diseases will be immense.

References

1   Battisto JR, Miller J: Immunological unresponsiveness produced in adult guinea pigs by parenteral introduction of minute quantities of hapten or protein antigen. Proc Soc Exp Biol Med 1962;111:111–115.
2   Asherson GL, Stone SH: Selective and specific inhibition of 24-hour skin reactions in the guinea pig. I. Immune deviation: description of the phenomenon and the effect of splenectomy. Immunology 1965;9:205–217.

3   Asherson GL: Selective and specific inhibition of 24-hour skin reactions in the guinea pig. II. The mechanism of immune deviation. Immunology 1966;10:179–186.

4   Dvorak HF, Billote JB, McCarthy JS, Flax MH: Immunological unresponsiveness in the adult guinea pig. I. Suppression of delayed hypersensitivity and antibody formation to protein antigens. J Immunol 1965;94:966–975.

5   Borel Y, Fauconnet M, Miescher PA: Selective suppression of delayed hypersensitivity by the induction of immunological tolerance. J Exp Med 1966;123:585–598.

6   Loewi G, Holborow EJ, Temple A: Inhibition of delayed hypersensitivity by preimmunization without complete adjuvant. Immunology 1966;10:339–347.

7   Crowle AJ, Hu CC: Split tolerance affecting delayed hypersensitivity and induced in mice by preimmunization with protein antigens in solution. Clin Exp Immunol 1966;1:323–335.

8   Crowle AJ, Hu CC: Studies on the induction and time course of repression of delayed hypersensitivity in the mouse by low and high doses of antigen. Clin Exp Immunol 1970;6:363–374.

9   Borel Y, David JR: In vitro studies of the suppression of delayed hypersensitivity by the induction of partial tolerance. J Exp Med 1970;131:603–610.

10 Janicki BW, Schechter GP, Schultz KE: Cellular reactivity to tuberculin in immune and serologically-unresponsive guinea pigs. J Immunol 1970;105:527–530.

11 Parish CR: Immune response to chemically modified flagellin. II. Evidence for a fundamental relationship between humoral and cell-mediated immunity. J Exp Med 1971;134:21–47.

12 Parish CR, Liew FY: Immune response to chemically modified flagellin. III. Enhanced cell-mediated immunity during high and low zone antibody tolerance to flagellin. J Exp Med 1972;135:298–311.

13 Liew FY, Parish CR: Lack of correlation between cell-mediated immunity to the carrier and the carrier-hapten helper effect. J Exp Med 1974;139:779–784.

14 Silver J, Benacerraf B: Dissociation of T cell helper function and delayed hypersensitivity. J Immunol 1974;113:1872–1875.

15 Marrack P, Kappler J: Antigen-specific and non-specific mediators of T cell/B cell co-operation. I. Evidence for their production by different T cells. J Immunol 1975;114:1116–1125.

16 Janeway CA: Cellular co-operation during in vivo antihapten antibody response. I. The effect of cell number on the response. J Immunol 1975;114:1394–1401.

17 Howard JG, Hale C, Liew FY: Immunological regulation of experimental cutaneous leishmaniasis. IV. Prophylactic effect of sublethal irradiation as a result of abrogation of suppressor T cell generation in mice genetically susceptible to Leishmania tropica. J Exp Med 1981;153:557–568.

18 Mitchell GF, Curtis JM, Handman E, McKenzie IFC: Cutaneous leishmaniasis in mice: disease patterns in reconstructed nude mice of several genotypes infected with Leishmania tropica. Aust J Exp Biol Med Sci 1980;58:521–532.

19 Liew FY, Hale C, Howard JG: Immunologic regulation of experimental cutaneous leishmaniasis. V. Characterisation of effector and specific suppressor T cells. J Immunol 1982;128:1917–1922.

20 Liew FY: Specific suppression of responses to Leishmania tropica by a cloned T-cell line. Nature 1982;305:630–632.

21 Mossman TR, Cherwinski H, Bond MW, Giedlin MA, Coffman RL: Two types of

murine helper T cell clone. I. Definition according to profiles of lymphokine activities and secreted proteins. J Immunol 1986;136:2348–2357.

22 Modabber F: The leishmaniasis; in Maurice J, Pearce AM (eds): Tropical Disease Research: A Global Partnership. 8th Programme Report, TDR. Geneva, World Health Organisation, 1987, pp 99–112.

23 Greenblatt CL: The present and future of vaccination for cutaneous leishmaniasis: in Mizrahi A, Hertman I, Klinberg MA (eds): New Developments with Human and Veterinary Vaccines. New York, Liss, 1980, pp 259–285.

24 Colle JH, Truffa-Bachi P, Chedid L, Modabber F: Lack of a general immunosuppression during visceral *Leishmania tropica* infection in BALB/c mice: augmented antibody response to thymus-independent antigens and polyclonal activation. J Immunol 1983;131:1492–1495.

25 Pearson RD, Steigbigel RT: Mechanism of lethal effect of human serum upon *Leishmania donovani*. J Immunol 1980;125:2195–2201.

26 Hertman R: Cytophilic and opsonic antibodies in visceral leishmaniasis in mice. Infect Immun 1980;28:585–593.

27 Olobo JO, Handman E, Curtis JM, Mitchell GF: Antibodies to *Leishmania tropica* promastigotes during infection in mice of various genotypes. Aust J Exp Biol Med Sci 1980;58:595–601.

28 Howard JG, Liew FY, Hale C, Nicklin S: Prophylactic immunisation against experimental leishmaniasis. II. Further characterisation of the protective immunity against fatal *L. tropica* infection induced by irradiated promastigotes. J Immunol 1984;132:450–455.

29 Lynch NR, Yarzabal L, Verdel O, Avila JL, Monzon H, Convit J: Delayed-type hypersensitivity and immunoglobulin E in American cutaneous leishmaniasis. Infect Immun 1982;38:877–881.

30 Lynch NR, Malave C, Turner KJ, Infante B: IgE antibody against surface antigens of Leishmania promastigotes in American cutaneous leishmaniasis. Parasite Immunol 1986;8:109–116.

31 Sacks DL, Scott PA, Asofsky A, Sher FA: Cutaneous leishmaniasis in anti-IgM-treated mice: Enhanced resistance due to functional depletion of a B cell-dependent T cell involved in the suppression pathway. J Immunol 1984;132:2072–2077.

32 Scott P, Natovitz P, Sher A: B lymphocytes are required for the generation of T cells that mediate healing of cutaneous leishmaniasis. J Immunol 1986;137:1017–1021.

33 Preston PM, Caster RL, Leuchars E, Davies AJS, Dumonde DC: Experimental cutaneous leishmaniasis. III. Effects of thymectomy on the course of infection of CBA mice with *Leishmania tropica*. Clin Exp Immunol 1972;10:337–357.

34 Mitchell GF, Curtis JM, Scollay RG, Handman E: Resistance and abrogation of resistance to cutaneous leishmaniasis in reconstituted BALB/c nude mice. Aust J Exp Biol Med Sci 1981;59:539–554.

35 Preston PM, Dumonde DC: Experimental cutaneous leishmaniasis. V. Protective immunity in subclinical and self-healing infection in the mouse. Clin Exp Immunol 1976;23:126–138.

36 Rezai HR, Farrell J, Soulsby EL: Immunological responses of *L. donovani* infection in mice and significance of T cell in resistance to experimental leishmaniasis. Clin Exp Immunol 1980;40:508–514.

37 Liew FY, Howard JG, Hale C: Prophylactic immunisation against experimental leishmaniasis. III. Protection against fatal *Leishmania tropica* infection induced by

irradiated promastigotes involves Lyt-1^+2^- T cells that do not mediate cutaneous DTH. J Immunol 1984;132:456–461.

38   Gorczynski RM, MacRae S: Analysis of subpopulations of glass adherent mouse skin cells controlling resistance/susceptibility to infection with *Leishmania tropica* and correlation with the development of independent proliferative signals to Lyt-1.2^+/ Lyt-2.1^+ T lymphocytes. Cell Immunol 1981;67:74–89.

39   Maul J, Behin R: Leishmaniasis; in Cohen S, Warren KS (eds): Immunology of Parasitic Infections. Oxford, Blackwell Scientific Publications, 1982, pp 299–355.

40   Titus RG, Ceredig R, Cerottini JC, Louis JA: Therapeutic effect of anti-L3T4 monoclonal antibody GK1.5 on cutaneous leishmaniasis in genetically susceptible BALB/c mice. J Immunol 1987;135:2108–2114.

41   Stern JJ, Oca MJ, Rubin BY, Anderson SL, Murray HW: Role of L3T4$^+$ and Lyt-2$^+$ cells in experimental visceral leishmaniasis. J Immunol 1988;140:3971–3977.

42   Smiter LE, Rodrigues M, Russell DG: Cytotoxic T cells and *Leishmania*-infected macrophages. J Exp Med 1991;174:459–506.

43   Rezai HR, Ardekali SM, Amirhakimi G, Kharazmi A: Immunological features of kala-azar. Am J Trop Med Hyg 1978;27:1079–1083.

44   Ho M, Koech DK, Iha DM, Bryceson ADM: Immunosuppression in Kenyan visceral leishmaniasis. Clin Exp Immunol 1983;51:207–214.

45   Carvalho EM, Teixeira R, Johnson WD: Cell-mediated immunity in American visceral leishmaniasis: reversible immunosuppression during acute infection. Infect Immun 1981;33:498–502.

46   Haldar JP, Ghose S, Saha KC, Ghose AC: Cell mediated immune response in Indian kala-azar and post kala-azar dermal leishmaniasis. Infect Immun 1983;42:702–707.

47   Reiner NE, Finke JH: Interleukin-2 deficiency in murine *Leishmania donovani* and its relationship to depressed spleen cell responses to phytohaemagglutinin. J Immunol 1983;131:1487–1491.

48   Scott PA, Farrell JP: Experimental cutaneous leishmaniasis. I. Nonspecific immuno-suppression in BALB/c mice infected with *Leishmania tropica*. J Immunol 1981;127:2395–2400.

49   Cillari E, Liew FY, Lelchuk R: Suppression of interleukin-2 production by macro-phages in susceptible BALB/c mice infected with *Leishmania major*. Infect Immun 1986;54:386–394.

50   Peterson EA, Neva FA, Barral A, Correa-Coronas R, Bogaert-Diaz H, Martinez D, Ward FE: Monocytes suppression of antigen-specific lymphocyte responses in diffuse cutaneous leishmaniasis patients from the Dominican Republic. J Immunol 1984;132:2603–2606.

51   Cillari E, Liew FY, Lo Campo P, Milano S, Mansueto S, Salerno A: Suppression of IL-2 production by cryopreserved peripheral blood mononuclear cells from patients with active visceral leishmaniasis in Sicily. J Immunol 1988;140:2721–2726.

52   Liew FY: Functional heterogeneity of CD4$^+$ T cells in leishmaniasis. Immunol Today 1989;10:40–45.

53   Scott P, Pearce G, Cheever AW, Coffman RL, Sher A: Role of cytokines and CD4 T cell subsets in the regulation of parasitic immunity and disease. Immunol Rev 1989;112:161–183.

54   Heinzel FP, Sadick MD, Holaday BJ, Coffman RL, Locksley RM: Reciprocal expression of interferon-γ or interleukin-4 during the resolution or progression of

murine leishmaniasis. Evidence for expansion of distinct helper T cell subsets. J Exp Med 1989;169:59–72.

55 Scott P, Natovitz P, Coffman RL, Pearce E, Sher A: Immunoregulation of cutaneous leishmaniasis. T cell lines that transfer protective immunity or exacerbation belong to different T helper subsets and respond to distinct parasite antigens. J Exp Med 1988;168:1675–1684.

56 Heinzel FP, Sadick MD, Mutha S, Locksley RM: Production of interferon-γ, interleukin-2, interleukin-4 and interleukin-10 by CD4⁺ lymphocytes in vivo during healing and progressive murine leishmaniasis. Proc Natl Acad Sci USA 1991;88:7011–7015.

57 Titus RG, Lima GC, Engers HD, Louis JA: Exacerbation of murine cutaneous leishmaniasis by adoptive transfer of parasite-specific helper T cell population capable of mediating *Leishmania major*-specific delayed-type hypersensitivity. J Immunol 1984;133:1594–1600.

58 Titus RG, Muller I, Kinsey P, Ceray A, Behin R, Zinkernagel RM, Louis J: Exacerbation of experimental murine cutaneous leishmaniasis with CD4⁺ *Leishmania major*-specific T cell lines or clones which secrete interferon-γ and mediate parasite-specific delayed-type hypersensitivity. Eur J Immunol 1991;21:559–567.

59 Dhaliwal JS, Liew FY, Cox FEG: Specific suppressor T cells for delayed type hypersensitivity in susceptible mice immunised against cutaneous leishmaniasis. Infect Immun 1985;49:417–423.

60 Liew FY, Dhaliwal JS: Distinctive cellular immunity to genetically susceptible BALB/c mice recovered from *Leishmania major* infection or after subcutaneous immunization with killed parasites. J Immunol 1987;138:4450–4456.

61 Kaye PM, Curry AJ, Blackwell JM: Differential production of Th₁ and Th₂-derived cytokines does not determine the genetically controlled or vaccine induced rate of cure in murine visceral leishmaniasis. J Immunol 1991;146:2763–2770.

62 Mason D, Powrie F: Memory CD4 T cells in man form two distinct subpopulations, defined by their expression of isoforms of the leukocytes common antigen CD45. Immunology 1990;70:427–433.

63 Romangnani S: Human Th₁ and Th₂ subsets: doubt no more. Immunol Today 1991;12:256–257.

64 Liew FY, Millott S, Li Y, Lelchuk R, Chan WL, Ziltener H: Macrophage activation by interferon-γ from host-protective T cells is inhibited by interleukin (IL)-3 and IL-4 produced by disease-promoting T cells in leishmaniasis. Eur J Immunol 1989;19: 1227–1232.

65 Lehn M, Weiser WY, Engelhorn S, Gillis S, Remold HG: IL-4 inhibits H₂O₂ production and anti-leishmanial capacity of human cultured monocytes mediated by IFN-gamma. J Immunol 1989;143:3020–3024.

66 Bogdan C, Stenger S, Rollinghoff M, Solbach W: Cytokine interaction in experimental cutaneous leishmaniasis. Interleukin-4 synergises with interferon-γ to activate murine macrophages for killing of *Leishmania major* amastigotes. Eur J Immunol 1991;21:327–333.

67 Fiorentino DF, Bond MW, Mosmann TR: Two types of mouse T helper cell. IV. Th2 clones secrete a factor that inhibits cytokine production by Th1 clones. J Exp Med 1989;170:2081–2095.

68 Fiorentino DF, Zlotnik A, Viera P, Mosmann TR, Howard M, Moore KW, O'Garra A: IL-10 acts on the antigen presenting cell to inhibit cytokine production by Th1 cells. J Immunol 1991;146:3444–3451.

69 de Wall Malefyt R, Haanen J, Spits H, Roncarolo M-G, te Velde A, Figdor C, Johnson K, Kastelein R, Yssel H, de Vries JE: Interleukin-10 (IL-10) and viral IL-10 strongly reduce antigen-specific human T cell proliferation by diminishing the antigen-presenting capacity of monocytes via down-regulation of class II major histocompatibility complex expression. J Exp Med 1991;174:915–924.

70 Mosmann TR, Moore KW: The role of IL-10 in cross-regulation of Th_1 and Th_2 responses; in Ash C, Gallagher RB (eds): Immunoparasitology Today. Cambridge, Elsevier Trends Journals, 1991, pp A49–A53.

71 Gajewski TF, Fitch FW: Anti-proliferative effect of IFN-γ in immune regulation. I. IFN-γ inhibits the proliferation of Th_2 but not Th_1 murine lymphocyte clones. J Immunol 1988;140:4245–4252.

72 Titus RG, Sherry B, Cerami A: Tumor necrosis factor plays a protective role in experimental murine cutaneous leishmaniasis. J Exp Med 1989;170:2097–2104.

73 Liew FY, Parkinson C, Millott S, Severn A, Carrier M: Tumour necrosis factor (TNFα) in leishmaniasis. I. TNFα mediates host protection against cutaneous leishmaniasis. Immunology 1990;69:570–573.

74 Liew FY, Li Y, Millott S: Tumour necrosis factor (TNFα) in leishmaniasis. II. TNFα-induced macrophage leishmanicidal activity is mediated by nitric oxide from L-arginine. Immunology 1990;71:556–559.

75 Bogdan C, Moll H, Solbach W, Rollinghoff M: Tumour necrosis factor-α in combination with IFN-γ, but not with IL-4 activates murine macrophages for elimination of Leishmania major amastigotes. Eur J Immunol 1990;20:1131–1135.

76 Janeway CA, Carding S, Jones B, Murray J, Portoles P, Rasmussen R, Rojo J, Saijawa K, West J, Bottomly K: CD4$^+$ T cells: specificity and function. Immunol Rev 1988;101:39–80.

77 Cillari E, Dieli M, Maltese E, Milano S, Salerno A, Liew FY: Enhancement of macrophage IL-1 production by Leishmania major infection in vitro and its inhibition by IFN-γ. J Immunol 1989;143:2001–2005.

78 Reiner NE, Ng W, McMaster WR: Parasite-accessory cell interactions in murine leishmaniasis. II. Leishmania donovani suppresses macrophage expression of class I and class II major histocompatibility complex gene products. J Immunol 1987;138: 1928–1932.

79 Howard JG, Nicklin S, Hale C, Liew FY: Prophylactic immunisation against experimental leishmaniasis. I. Protection induced in mice genetically vulnerable to fatal Leishmania tropica infection. J Immunol 1982;129:2206–2211.

80 Liew FY, Hale C, Howard JG: Prophylactic immunisation against experimental leishmaniasis. IV. Subcutaneous immunisation prevents the induction of protective immunity against fatal Leishmania major infection. J Immunol 1985;135:2095–2101.

81 Liew FY, Millott SM, Schmidt JA: A repetitive peptide of leishmania can activate Th_2 cells and enhance disease progression. J Exp Med 1990;172:1359–1365.

82 Yang DM, Rogers MV, Liew FY: Identification and characterisation of host-protective T cell epitopes of a major surface glycoprotein (gp63) from Leishmania major. Immunology 1990;72:3–9.

83 Bouvier J, Bordier C, Vogel H, Reichelt R, Etges R: Characterisation of the promastigote surface protease of leishmania as a membrane-bound zinc endopeptidase. Mol Biochem Parasitol 1989;37:235–246.

84 Chaudhuri G, Chandhuri M, Pan A, Chang K-P: Surface acid proteinase (gp63) of Leishmania mexicana: a metalloenzyme capable of protecting liposome-encapsu-

lated proteins from phagolysosomal degradation by macrophages. J Biol Chem 1989; 264:7483–7489.

85 Kink JA, Chang K-P: Biological and biochemical characterization of tunicamycin-resistant *Leishmania mexicana*: Mechanism of drug resistance and virulence. Infect Immun 1987;55:1692–1700.

86 Liew FY, Li Y, Yang DM, Severn A, Cox FEG: TNFα reverses the disease exacerbating effect of subcutaneous immunisation against murine cutaneous leishmaniasis. Immunology 1991;74:304–309.

87 Moncada S, Palmer RMJ, Higgs EA: Nitric oxide: Physiology, pathology and pharmacology. Pharmacol Rev 1991;43:109–142.

88 Green SJ, Meltzer MS, Hibbs JB Jr, Nacy CA: Activated macrophages destroy intracellular *Leishmania major* amastigotes by an L-arginine-dependent killing mechanism. J Immunol 1990;144:278–283.

89 Liew FY, Millott S, Parkinson C, Palmer RMJ, Moncada S: Macrophage killing of *Leishmania* parasite in vivo is mediated by nitric oxide from L-arginine. J Immunol 1990;144:4793–4797.

90 Mauel J, Ransijn A, Buchmuller-Rouiller Y: Killing of *Leishmania* parasites in activated murine macrophages is based on an L-arginine-dependent process that produces nitrogen derivatives. J Leukoc Biol 1990;49:73.

91 Liew FY, Li Y, Moss D, Parkinson C, Rogers MV, Moncada S: Resistance to *Leishmania major* infection correlates with the induction of nitric oxide synthase in murine macrophages. Eur J Immunol 1991;21:3009–3014.

92 Liew FY, Li Y, Severn A, Millott S, Schmidt J, Salter M, Moncada S: A possible novel pathway of regulation by murine Th_2 cells of a Th_1 cell activity via the modulation of the induction of nitric oxide synthase on macrophages. Eur J Immunol 1991;21:2489–2494.

93 Weiser WY, Temple PA, Witek-Giannotti JS, Remold HG, Clark C, David JR: Molecular cloning of a cDNA encoding a human macrophage migration inhibition factor. Proc Natl Acad Sci USA 1989;86:7522–7526.

94 Weiser WY, Pozzi L-AM, David JR: Human recombinant migration inhibitory factor activates human macrophages to kill *Leishmania donovani*. J Immunol 1991;147:2006–2011.

95 Ding A, Nathan CF, Graycar J, Derynct R, Stuehr DJ, Srimal S: Macrophage deactivating factor and transforming growth factor-β1, -β2 and -β3 inhibit induction of macrophage nitrogen oxide synthesis by IFN-γ. J Immunol 1990;145:940–944.

96 Liew FY, Cox FEG: Non-specific defence mechanism: the role of nitric oxide; in Ash C, Gallagher R (eds): Immunoparasitology Today. Cambridge, Elsevier Trends Journals, 1991, pp 17–21.

97 Bredt DS, Hwang PM, Glatt CE, Lowenstein C, Reed RR, Snyder SH: Cloned and expressed nitric oxide synthase structurally resembles cytochrome P-450 reductase. Nature 1991;351:714–718.

F.Y. Liew, PhD, Department of Immunology, University of Glasgow,
Western Infirmary, Glasgow, G11 6NT (UK)

Coffman RL (ed): Regulation and Functional Significance of T-Cell Subsets.
Chem Immunol. Basel, Karger, 1992, vol 54, pp 136–165

Biological Role of Helper T-Cell Subsets in Helminth Infections

Christopher L. King[a], Thomas B. Nutman[b]

[a]Division of Geographic Medicine, Case Western Reserve University,
Cleveland, Ohio, and [b]Laboratory of Parasitic Diseases,
National Institutes of Health, Bethesda, Md., USA

Introduction

The hallmarks of helminth parasite infections are elevated serum IgE levels, peripheral and tissue eosinophilia and the participation of inflammatory mediator-rich basophils and mast cells – each of which, in part, mediates immediate hypersensitivity responses. These responses have been clearly implicated in the pathogenesis of allergic diseases; however, their role in parasitic infections is less clear. Although these responses can induce pathologic reactions, they may also play an important role in modulating the host response to the parasite and ultimately in inducing protective immunity.

Cytokines produced by CD4$^+$ T cells have been shown to control IgE production, eosinophilia and mastocytosis. For IgE, production is regulated by IL-4 and IFN-γ [1–3], and IL-5 has been shown to be the major eosinophilopoietin and also capable of supporting the growth of basophils [4, 5]. Populations of CD4$^+$ T cells (termed Th1 and Th2) have been identified based on their mutually exclusive production of IL-2 and IFN-γ (Th1) or IL-4 and IL-5 (Th2 [6, 7]). In the murine system, Th1 cells mediate several functions associated with local inflammatory reactions and cytotoxicity whereas Th2 cells are more effective at providing help for antibody production [8]. It should be emphasized that some cytokines, such as IL-3, GM-CSF, and TNF are produced by both Th1 and Th2 cells; thus the polarization in specific cytokine production between these two subsets is not always distinct. In addition, there has been recently described CD4$^+$ T cells capable of producing IL-4 and IL-5 as well as IFN-γ and IL-2 which have been designated Th0 [9]. It has been hypothesized that as antigen-specific

T cells are generated and expanded they progress through an ordered sequence of cytokine production including Th0 cells which likely differentiate further to Th1 or Th2 cells with repeated stimulation. This may explain why some seemingly disparate immunological responses often accompany each other. Ultimately a generally stable relationship between the Th1 and Th2 CD4$^+$ populations is generated, in part through the production of cross-regulatory cytokines (e.g. IL-10 and IFN-γ [10]). It is a shift to a Th2-type of cytokine pattern that is felt to mediate those responses associated with parasite infection. How a predominantly Th2 pattern is generated and the functional significance of these responses in regulating the host-parasite relationship remain open questions.

This review will attempt to summarize the evidence linking specific cytokines and CD4$^+$ T-cell subset to the generation of immediate hypersensitive responses in humans. Moreover, we will focus on the biological role of the helminth-induced Th2 cell responses in both murine and human systems and discuss some of the hypotheses generated to understand how these responses are produced.

Cytokine Regulation of the Immune Responses –
IgE, Eosinophilia, and Mastocytosis

IgE
A clear association between IL-4 and the production of IgE in murine systems has been demonstrated [11]. Perhaps the most convincing evidence exists in experimental models of murine helminth infections. Mice infected with *Schistosoma mansoni, Nippostrongylus brasiliensis* or *Heligmosomoides polygyrus* generate marked polyclonal IgE and IgG1 responses that are almost totally abrogated by injection of anti-IL-4 and/or anti-IL-4 receptor monoclonal antibodies.

The requirement for IL-4 in the generation of increased IgE production in humans is becoming increasingly evident, although the data is more correlative in nature. In vitro studies have demonstrated that recombinant IL-4 added to normal human PBMC induces IgE production that can be inhibited by the addition of rIFN-γ [2, 12, 13]. In addition, supernatants generated from T-cell clones obtained from patients with helminth infections or atopic disorders generally produce high levels of IL-4 and little IFN-γ that can, in turn, stimulate IgE production by normal PBMC [14–17]. Closer to the natural state are studies demonstrating that parasite antigen-induced polyclonal IgE production by fresh PBMC obtained from patients with active helminth infections is reciprocally regulated by IL-4 and IFN-γ [18]. The IgE response can be completely blocked by neutralizing antibodies to IL-4 and augmented by anti-IFN-γ antibodies, demonstrating that endogenously produced cytokines likely regulate IgE production in natural infections. Similar studies have also indicated a role of IL-4 in the induction of defined allergen-driven IgE production in vitro by PBMC from atopic patients [19].

Direct detection of increased IL-4 production (and correspondingly diminished IFN-γ production) in disorders associated with elevated serum IgE levels has been difficult to demonstrate in part because of the ephemeral nature of IL-4 (also a T-cell growth factor) and the low frequency of cells producing it. Because helminth infections evoke IgE and eosinophilic responses that quantitatively are at least 10–100 times greater than those seen in allergic disorders, the study of their regulation in these parasite infections has, in many ways, been more approachable experimentally. For example, mitogenic stimulation of PBMC obtained from 18 helminth-infected patients (with correspondingly increased serum IgE levels compared to 9 normal individuals), showed a 4.5-fold greater IL-4 production in culture supernatants [20]. The specificity of this response was further demonstrated when parasite antigen-driven IL-4 production in PBMC culture supernatants from 14 helminth-infected patients was readily detectable; in contrast, nonparasite antigen failed to induce measurable IL-4 production in the same individuals. More importantly, parasite antigen-induced IL-4 production significantly correlated with serum parasite-specific IgE levels, thus linking IL-4 production with the generation of IgE response in vivo [C.L. King et al., unpubl. observations]. Conversely, IFN-α and IFN-γ inhibit IgE production in vivo based on studies in which these cytokines were administered directly to patients with disorders of excessive IgE production (e.g. hyper-IgE-recurrent infection syndrome and atopic dermatitis) and suppressed both serum IgE levels as well as spontaneous in vitro IgE production by their PBMC obtained during therapy [21, 22].

Eosinophilia

In the same murine models of helminth infections described above, a direct relationship between IL-5 and the generation of eosinophilia has also been demonstrated [23–25]. Furthermore, direct administration of neutralizing monoclonal antibodies to IL-5 completely blocks peripheral and tissue eosinophilia in mice infected with *S. mansoni, N. brasiliensis* or *Stron-*

gyloides venezuelensis [23, 26, 27]. In humans, IL-5 has been implicated in mediating eosinophilia in vitro both in marrow cultures [4] and in cross-sectional studies of eosinophilic individuals [28]. Although other cytokines, particularly IL-3 and GM-CSF, have been shown to stimulate eosinophilo-poiesis directly in vitro and to induce eosinophilia in vivo [29], only IL-5 appears to be eosinophil- (and probably basophil)-specific in its colony-stimulating activity [30, 31]. The role of IL-5 in the induction of eosinophilia in humans is strongly suggested by the finding of transiently elevated levels of IL-5 just prior to the development of posttreatment eosinophilia in patients infected with onchocerciasis [32]. Furthermore, IL-5 has also been detected in the serum from patients with chronic schistosomiasis and markedly elevated peripheral eosinophilia [33]. Other noninfectious causes of eosinophilia in humans have been also causally linked to increased IL-5 production [34, 35]. On the other hand, the cytokine IFN-α appears to antagonize eosinophilia in *N. brasiliensis* mice [11] and possibly in humans where it has been demonstrated to suppress eosinophilia in 2 patients with the idiopathic hypereosinophilic syndrome [36, 37].

Mastocytosis

Mast cells appear to play an important role in the inflammatory reactions against invasive helminth parasites in that these cells are generally present in elevated numbers in parasitized organs [reviewed in 38], have high affinity Fc-ε receptors that are sensitized with specific antiparasite IgE, and, because they are continuously exposed to parasite antigens, have easily triggered mediators. The presence of the mast cells may have been an evolutionary adaptation to infection with parasitic helminths in that (unlike their role in allergic diseases) they somehow confer a biological advantage to the human host [38–43]. Indeed it has been postulated that mediators specifically released by sensitized mast cells contribute to: (1) the recruitment and activation of effector eosinophils [44, 45]; (2) an increase in local concentrations of antibody and complement [46, 47], and (3) in the intestinal tract, mucus hypersecretion [48–51] and increased peristalsis [52] which could interfere with the attachment process of certain parasites that would be necessary for the establishment of chronic infection. The induction of mastocytosis has been linked to several cytokines. The first cytokine demonstrated to be important in this process was IL-3 which was shown to stimulate mast cell production and differentiation in both murine and human bone marrow cultures [53, 54]. The essential role of IL-3 and IL-4 in the induction of intestinal mast cell hyperplasia in mice infected by the intestinal nematode

N. brasiliensis has also been recently demonstrated [55]. Administration of neutralizing antibodies to either IL-3 or IL-4 blocks mastocytosis by approximately 50%; treatment with both antibodies can suppress the mucosal mast cell response by 85–90%. The cytokines IL-9 and IL-10 along with stem cell factor have also been demonstrated to be important in the generation of mastocytosis in vitro [56–58]; however, their role in conjunction with IL-3 and IL-4 for the induction of mastocytosis in vivo remains to be elucidated.

Biological Role of CD4⁺ Subsets in Animal Models

The notion that specific cytokines can regulate distinct immune effector systems has been extended to the concept that certain CD4⁺ populations generate characteristic patterns of cytokines, e.g. Th1 and Th2 subsets [7]. The functional significance for the existence of these subsets, however, is not well understood. Thus, studies in laboratory models in which these responses are most pronounced may be those best utilized to understand their importance. Helminth infections, for example, consistently generate pronounced Th2 responses as the infection becomes chronic. Studies in *S. mansoni* infection in naive mice show an early and predominant Th1 response; at the onset of egg production by the adult worms, however, the response switches to a predominantly Th2 response (IL-4, IL-5 and IL-10 synthesis) with a concomitant decrease in IL-2 and IFN-γ production [59, 60]. This drop in IL-2 and IFN-γ production appears to result from active suppression of Th1 cells by IL-10 [61]. In this example, certain stage-specific antigens appear to be responsible for the induction of the Th2 response, rather than the chronicity of the infection. The exact kinetics of the induction of this response has not been clearly defined in the natural infection; however, it appears to remain suppressed for up to 22 weeks after primary infection [62]. In view of earlier studies that the granulomatous reaction around *S. mansoni* eggs in the liver becomes less severe as the infection proceeds [63] suggests that the inhibition of IL-2 and IFN-γ (and possibly TNF) may serve to suppress the immunopathologic responses to chronic infection.

The generation of Th2 responses in helminth infections with the associated immediate hypersensitivity responses may not only function to modulate the immunopathology to infection, but also serve to induce protective immunity. In vitro studies, particularly with *S. mansoni* infec-

Table 1. Animal models of helminth infections

Parasite	CD4+ response	Effect on immunity (in vivo mAb treatment)				Ref.
		anti-IL-4 (anti-IL-4R)	anti-IL-5	anti-IFN-γ	anti-IgE	
Schistosoma mansoni	Th1-Th2	$-^1$	−	$+^2$	ND	25, 26
Nippostrongylus brasiliensis	Th2	−	−	ND	ND	11, 24
Heligmosomoides polygyrus	Th2	+	−	ND	−	67
Trichinella spiralis	Th1-Th2	ND	ND	ND	+	120
Strongyloides venezuelensis	Th2	ND	+	ND	ND	27

[1]No effect on immunity.
[2]Ablates protective immunity.

tions, indicate that IgE and eosinophils are crucial for killing the parasites [64]. These observations were partially based on studies showing that eliminating eosinophils or IgE in vivo with specific antibodies eliminates resistance to this and other helminth parasites [65, 66, 121]. More recently, however, inhibition of immediate hypersensitivity responses by specific anticytokine antibodies in helminth-infected mice have produced conflicting results (table 1). Consistently, anti-IL-5 mAb abrogated both peripheral and tissue eosinophilia, but failed to block immunity to reinfection with *S. mansoni, N. brasiliensis,* and *H. polygyrus* [11, 26, 67]. In contrast, however, neutralizing antibodies to IL-5 or IL-5 receptor blocked peripheral and tissue eosinophilia to *S. venezuelensis* infection and also abolished expression of protective immunity to challenge infection, although primary infection was not affected [27]. The obvious difference between these studies was the use of different anti-IL-5 mAb. The same antibody (TRFK-5) was used in the three studies in which protective immunity was not effected; however, a different monoclonal (NC17) was used in the *S. venezuelensis* experiments. Although both antibodies inhibited the eosinophilic response, it is possible that already existing eosinophils may be sufficient to effect larval destruction and the antibody NC17 may have greater affinity for IL-5 and alter the ability of the remaining eosinophils to respond.

Antibodies to IL-4 and/or IL-4R failed to inhibit immunity to reinfection in *S. mansoni* and *N. brasiliensis* models; however, protective immunity to *H. polygyrus* was ablated. Unlike the other two helminth infections, *H. polygyrus* is a completely enteric infection, although considerable tissue inflammation occurs after the parasitic third-stage larvae penetrates the intestinal mucosa. Thus, the primary effector mechanisms of protective immunity in *H. polygyrus* infection may occur in the intestinal lumen in association with intestinal mastocytosis – a situation where IgE may be a more critical effector molecule.

In contrast, antibodies to IFN-γ in murine *S. mansoni* infections can partially inhibit immunity when administered 1 day before and weekly after challenge infection in mice immunized with irradiated cercariae [25]. This immunity can be completely abrogated if the anti-IFN-γ antibody is administered 5 days after challenge infection when the larvae are migrating through the lung [R.A. Wilson, pers. commun.]. Lung biopsies performed at that time show marked pulmonary inflammation and extravasation of larvae into the lung alveoli. This suggests that the effector mechanisms needed to eliminate this stage of the parasite may be nonspecific. Overall, these studies suggest that the Th2 response engendered by helminth infections in rodents may be more important in modulating immunopathologic events than generating effector responses against worms.

Evidence for CD4+ Subsets in Humans

T-Cell Clones

The use of T-cell clones has been instrumental for studying the T-cell-dependent regulatory aspects of parasitic infections as it has been in the other infections and in allergic disease. Work with allergen- and parasite-specific T-cell clones has shown unequivocally that parasite- and allergen-specific T-cell responses are MHC class II restricted [68–70]. They have also been used as tools to explore the role of cytokines important in mediating immediate hypersensitivity responses; in fact, in humans they very clearly provided the initial evidence of activities that are now ascribed to IL-4 and IL-5. Because many of these studies were done with allergen specific clones that appear to be similar to those found in parasite infections, descriptions of some of these T-cell clones will also be discussed.

This section will detail those studies that utilized human T-cell clones derived from patients with parasitic helminth infection and allergic diseases and contrast them to clones derived from patients with chronic bacterial and protozoal infections. It is divided into two major parts – that dealing with T-cell clones without defined antigen specificity and another pertaining to T-cell clones in which the antigen specificity is known.

T-Cell Clones without Defined Specificity
Virally Transformed T-Cell Clones. Relatively little work has been performed using virally transformed T cells from humans with allergic diseases or parasitic infections. Nevertheless, using HTLV-1, T cells from patients with tropical pulmonary eosinophilia have been transformed, cloned and the cytokine profiles studied. The T cells fell into primarily three patterns: a Th1-like cytokine profile (13%), a Th2-like cytokine profile (33%) and a profile where there was secretion of most of the measurable T-cell-derived cytokines (64%).

In an additional study, there was evidence that the nature of the stimulus needed to activate the T cell prior to HTLV-1 transformation determined the ultimate characteristic of the transformed T-cell clone. Indeed, when a parasite antigen was used to stimulate the T cells prior to transformation and cloning there was an overrepresentation of T-cell clones capable of making IL-4 and IL-5. When tetanus toxoid (TT) was used as the activating agent in the same individual, the majority of the clones were incapable of producing either of these two cytokines, but did produce IL-2 and IFN-γ. When the mitogen PHA was utilized, the cells for the most part did not produce Th2-like cytokines; rather, they either produced exclusively IL-2 and IFN-γ or they produced most of the measurable T-cell-derived cytokines.

Mitogen-Driven T-Cell Clones. T-cell clones derived without regard to specificity have been generated to examine the role of the T cell in the regulation of IgE and eosinophilia in allergic diseases, parasitic infection and several unusual disorders of either IgE or eosinophil regulation [17, 71–74]. Specifically, large panels of T-cell clones have been derived using mitogens and IL-2 (and in some cases IL-4) using cells from patients with disorders of immediate hypersensitivity to examine the T cells' functional capacities. For example, T-cell clones from normal individuals were examined for their ability to both produce cytokines (including IL-4) and to provide helper function for IgE production in vitro systems [75, 76]. While the T-cell clones studied had patterns of cytokine production that varied (including many that made most of the measurable cytokines), there was found to be a relationship between the ability to produce IL-4 and the ability to provide help for IgE. Indeed those T-cell clones making both IL-4 and IFN-γ (compared to those making only IL-4) were less capable of helping B cells to produce IgE. These observations were extended to an examination of T-cell clones derived from lymphocytes of atopic patients and those with parasitic helminth infections (filariasis, toxocariasis, ascariasis) [17]. In general, there was a considerably larger percentage of T cells producing IL-4 and a smaller percentage of T cells producing IFN-γ in these patients than in normal, healthy control individuals. These findings suggested that in disorders associated with increased IgE production, the ability of T cells to generate preferentially IL-4 and not IFN-γ was the most important determinant for expression of disease.

Because cytokines more often act locally rather than at a distance (as hormones characteristically do) an examination of T-cell clones derived from specimens obtained from inflammatory foci has provided insights into the pathogenesis of some conditions. Indeed, when mitogen-driven T-cell clones were derived from conjunctival biopsies of patients suffering from vernal conjunctivitis – a seasonal disorder of conjunctival inflammation associated with papillar hypertrophy and cobblestoning along with elevated IgE (in serum and tears) – 92% of these were capable of producing IL-4 compared to 42% of the clones derived from the peripheral blood of the same patients [77]. This suggests that local overproduction of IL-4 and diminished production of IFN-γ may be responsible for the pathologic alterations seen [76]. This observation has been given further support by IL-4 transgenic mice in which eye pathology similar to that observed in vernal conjunctivitis has been described [78].

In a related study of T-cell clones derived from infiltrating T cells in autoimmune thyroiditis – a disease not characterized by immediate hypersensitivity reactions – the overwhelming majority (> 80%) were producing IFN-γ and not IL-4 [79]. Taken together, these findings suggest that the nature of the local T-cell populations may determine the expression of disease at a given anatomical site.

Antigen-Specific T-Cell Clones. While much has been learned using nonspecific agents and IL-2 to expand and study T cells derived from patients with allergic disorders and parasitic diseases, the ability to generate T-cell clones with a defined specificity allowed for a more critical examination of the interaction between antigens and T cells. In general, the antigen-specific

Table 2. Allergen and parasite T-cell clones

Diagnosis	Antigen[1]	n	IL-2	IFN-γ	IL-4	IL-5	Help for IgE	Ref.
Atopic dermatitis	Dp	9	66%	0	100%	ND	+	69, 82
Atopic dermatitis	TT/CA	7	100%	100%	43%	ND	ND	69, 82
Allergic asthma	Dp	3	100%	0	100%	ND	+	69, 82
Normal	Dp	7	86%	100%	27%	ND	–	69, 82
Allergic asthma	TT	28	ND	86%	93%	40%	+/–	83
Allergic asthma	Dp	43	ND	56%	95%	100%	+	83
Allergic rhinitis	PPD	10	ND	100%	90%	25%	+	83
Allergic rhinitis	Lol1	18	ND	90%	50%	55%	+/–	83
Normal	PPD	60	100%	100%	23%	23%	–	84
Normal	Toxocara	69	26%	26%	88%	88%	+	84
Loiasis	Filarial Ag	8	20%	20%	100%	100%	+	14
Onchocerciasis	*O. volvulus*	32	ND	25%	87%	100%	80%	Nutman, unpubl.
Leprosy	Mycob. hsp	9	ND	100%	44%	22%	ND	85
Leprosy	TT	7	ND	100%	100%	29%	ND	85
Lyme disease	Bb	14	ND	100%	0%	0%	ND	86

[1]Dp = *Dermatophagoides pteronyssinus*; TT = tetanus toxoid; CA = *Candida albicans*; Lol1 = *Lolium perenne* 1; PPD = mycobacterial purified protein derivative; Mycob. hsp = mycobacterial heat-shock protein; Bb = *Borrelia burgdorferi*.

T cells that have been studied are CD4[+] and recognize the antigens in an MHC class II-restricted manner. Initially, these T cells were used to study the T-cell arm of the immune response in the regulation of IgE synthesis in vitro [17, 69, 80–82]. More recently, these T cells have been studied to demonstrate the existence of Th1- and Th2-like cells in humans [82, 83]. A summary of some of the recent findings (published and unpublished) can be found in table 2. In sensitized individuals, allergens and/or parasite antigens select for T-cell clones that fall more into the Th2-like pattern of cytokine secretion than do antigen-driven T-cell clones obtained from patients with chronic bacterial infections [84, 85]. These results suggest bacterial antigens preferentially select IL-2 and IFN-γ producing T-cell clones.

Allergen-Specific. The factors responsible for the expression of allergic disease are complex but very clearly require a genetic predisposition and the ability to generate an IgE response that is allergen-specific. The induction of this IgE response was demonstrated to be T-cell-dependent, and this observation provided the framework to begin to examine the T-cell-allergen interaction. With the ability to generate antigen-specific human T-cell clones, panels of allergen-specific T-cell clones were first examined for their ability to provide the necessary 'help' for B-cell production of IgE. In a number of studies [19, 69, 82–84], using allergen-specific T-cell clones, the provision of help for IgE production could easily be demonstrated with autologous B cells in coculture experiments. When cytokine determinations were made of these T-cell clones, IL-4 was demonstrated to be the cytokine most likely responsible for the induction of the IgE response.

These allergen-specific T-cell clones provided additional support for the hypothesis that there is a genetic basis for allergic diseases. In three separate studies using Dermatophagoides as the stimulating antigen for the generation of T-cell clones from cells of either allergic or nonallergic individuals [19, 69, 82], it was clear that nonallergic individuals had a repertoire of dust mite-specific T cells that produced significant amounts of IFNγ but little IL-4. In contrast, close to 100% of Dermatophagoides-specific T-cell clones from atopic individuals produced IL-4 suggesting that not only was appropriate allergen sensitization necessary but so was having the underlying capacity to mount an 'allergic'-type immune response.

Equally important were the use of these allergen-specific T-cell clones to demonstrate unequivocally the presence in humans of Th1- and Th2-type cells [83, 84]. Using a panel of allergen- and bacterial antigen-specific T cells derived from the same donors, it was shown that antigen-specific Th1 (only IL-2 and IFN-γ producing) and Th2 (exclusively IL-4 and IL-5) T cells could be demonstrated. Moreover, allergens such as *Lolium perenne* group 1 (Lol1) or *Dermatophagoides pteronyssinus* were more likely to give rise to Th2-like clones whereas TT or purified protein derivative (PPD) generated Th1-like clones [83, 84]. That most of the allergen-specific helper T cells were able to produce high levels of IL-4 and IL-5 but not IFN-γ certainly provides a theory for why allergens induce elevated levels of IgE and peripheral blood eosinophils.

Parasite-Specific. Because parasitic helminth infections often stimulate exuberant immediate hypersensitivity responses, they have provided good experimental models. For instance, a soluble factor (now determined to be IL-4) derived from filarial parasite-specific T-cell clones and shown to be capable of inducing IgE production in normal B cells [14] was first noted in the early 1980s, well before such factors could be demonstrated in allergen-specific systems. More importantly, the use of parasite-specific T-cell clones has provided insights into the functions of the parasite antigen-specific T cell that very much parallel those found with allergen-specific T cells in atopic disorders. As seen in table 2, a large number of filarial-specific T-cell clones derived from patients with active loiasis or onchocerciasis (with associated elevations of IgE and peripheral blood eosinophils) had cytokine profiles that were Th2 in nature. In a similar but more broad-reaching study [84], it was found that when Toxocara antigen was used to generate antigen-specific T cells from a normal individual, this antigen preferentially induced cells capable of making IL-4 and IL-5, whereas PPD preferentially induced IL-2 and IFN-γ from the same individual. Work is currently in progress using these T-cell clones to focus in on the nature of the antigenic epitopes responsible for the divergent cell types seen, the role of antigen processing in this process, and the localization of these specific cytokine-producing T cells in the lesions associated with helminth parasitic infection.

Bacterial Antigen-Specific. In contrast to allergen or parasite antigen-specific T-cell clones that preferentially produce IL-4 and IL-5, T-cell clones obtained from patients with chronic bacterial infections produce primarily IFN-γ and/or IL-2 [85, 86]. In a study of patients with chronic lyme arthritis, *Borrelia burgdorferi* antigens were used to generate antigen-specific T cells that produced exclusively IFN-γ, GM-CSF, and TNF-α, but not IL-4 or IL-5, a pattern of cytokine production resembling Th1 cells [85] (see table 2). In a similar study of patients with tuberculoid leprosy, mycobacterial heat-shock proteins induced T-cell clones that produced very high levels of IFN-γ [85]. Although almost half of these clones secreted low levels of IL-4 and IL-5, the ratio of IFN-γ to IL-4 producing clones was much greater than that of clones reactive with nonmycobacterial antigens. The distinct dichotomy of cytokine patterns of T-cell clones obtained from chronic bacterial infections in contrast to those of parasite- or allergen-specific T-cell clones suggests that it is less likely to be the chronicity of the infection that determines the nature of cytokine production, but rather other factors such as the nature of the antigen or mechanism of antigen presentation (see below). Indeed, one of the control antigens used to generate clones from the same individuals both in the chronic bacterial studies and in some of the parasite and allergen studies was TT to which most North Americans have been immunized. Interestingly, the cytokine profile of TT-specific clones was usually mixed, resembling that of the Th0 cells. This difference might result from the fact that antigen-reactive T-cell clones obtained from individuals with chronic antigen stimulation (e.g.

chronic infections or allergies) polarize Th subset responses, whereas infrequent immunization with soluble proteins (TT) generates memory cells (e.g. CD4+, CD45RO+) with a predominantly Th0 phenotype [87, 88].

Direct Determination of Th Subsets by Immunoplaque Assays

Although the studies of T-cell clones provide compelling evidence for the presence of T-cell subsets in humans, these clones may not be representative of the entire T-cell repertoire in vivo. Furthermore, conditions used to generate clones from a single precursor in vitro may introduce a selection bias. To ascertain directly the frequency of individual cytokine-producing T cells, studies have been performed on helminth-infected patients using cytokine assays of individual cells [20, 89]. These studies have initially focused on recently infected North Americans traveling or working in areas endemic for helminth infections. In one such study, PBMC were obtained from 18 infected and 9 normal individuals and the frequency of IL-4-, IL-5-, GM-CSF- and IFN-γ-secreting CD4+ cells in response to mitogenic stimulation was compared between the two groups [20]. There was a 5-fold increase in the frequency of IL-5-secreting lymphocytes and a 2.5-fold increase in the frequency of IL-4-secreting lymphocytes among helminth-infected compared to normal individuals. No difference in the frequency of GM-CSF- and IFN-γ-producing cells between the two groups was observed, however, indicating a preferential expansion of Th2-type cells associated with parasite infections. Furthermore, there was a direct correlation between the amount of IL-4 and IL-5 produced in culture supernatants and the frequency of T cells producing these cytokines, indicating that the regulation of IL-4 and IL-5 may be linked.

Although the ability to measure antigen-driven IL-4- and IL-5-secreting T cells has been more difficult, the preferential expansion of parasite antigen-driven Th2 responses compared to nonparasite antigens is more dramatic than that observed to mitogens. Table 3 shows data from 4 North Americans with acute helminth infections, but who had been boosted with TT (within the past 7 years). Freshly isolated PBMC were enriched for CD4+ cells and stimulated in the presence of autologous adherent cells and the appropriate species-specific preparation of parasite antigens. The frequency of parasite antigen-reactive CD4+ lymphocytes secreting IL-4 ranged between ~ 1 in 800 to 1 in 2,000 whereas TT induced little Th2 responses. There was, however, detectable parasite antigen-driven IFN-γ producing T cells of comparable frequency to that induced by TT. As expected, parasite antigens failed to stimulate any IL-4-secreting lymphocytes in 5 normal individuals and only a

Table 3. Frequency of antigen-secreting CD4$^+$ cells from patients with acute helminth infections

Infection	Frequency of antigen-reactive CD4$^+$ T cells			
	IL-4		IFN-γ	
	parasite antigen	TT	parasite antigen	TT
Onchocerciasis	1/1,000	$<1/10^5$	1/370	1/526
Strongyloidiasis	1/794	$<1/10^5$	1/714	1/625
Lymphatic filariasis	1/1,841	1/12,500	1/1,429	1/769
Schistosomiasis	1/2,083	1/24,390	1/588	1/286

rare antigen-driven IFN-γ response (data not shown). It is of particular interest that these helminth-infected individuals also generated strong Th1 responses to parasite antigens. Whether this relates to the complexity of the antigenic mixture used (e.g. some antigens generate a Th2 response whereas others generate a Th1 response) or to the fact that the patients studied had acute rather than chronic infections remains to be evaluated. These results do support the concept that a functional dichotomy of CD4$^+$ T cells exists in humans in vivo.

The above studies are from nonendemic individuals acutely infected by their first exposure to these infections as adults. To understand the nature of the immune response in the more usual state, that of individuals from endemic areas with chronic infections, the frequencies of antigen-reactive lymphocytes were determined using ELISPOT among an endemic population for *Wuchereria bancrofti* in India. Twenty adults of both sexes manifesting a range of clinical expressions of the disease were studied. As shown in figure 1, parasite antigens preferentially induced a strong Th2 response, in contrast to PPD (an antigen to which the population has been sensitized). As observed in the acute helminth infections (table 3), parasite antigen appeared to also stimulate a Th1 response. The mixed nature of this Th1 and Th2 response in chronic lymphatic filariasis is striking in contrast to that observed with mycobacterial infections and murine models of helminth infections. As suggested before, the mixed nature of the parasite antigens with some generating a Th2 response (e.g. allergens) and others inducing predominantly a Th1 response is possible, but PPD, also a heterogenesis antigen mixture, was fairly uniform in inducing a Th1 response.

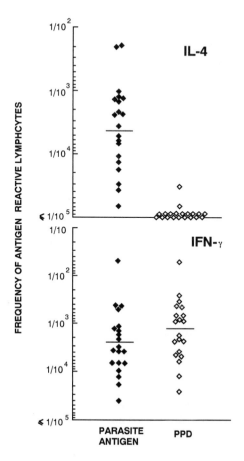

Fig. 1. The frequency of antigen-reactive T lymphocytes secreting IL-4 or IFN-γ from patients with lymphatic filariasis in response to crude parasite antigens (dark diamonds) or to mycobacterial purified protein derivative (PPD; open diamonds). Each point represents 1 individual and the horizontal lines represent geometric means.

Functional Significance CD4+ Subpopulations in
Human Helminth Infections

Filariasis

The study of lymphatic filariasis provides a particularly good system to examine the question of the biologic importance of the relationship of antigen-reactive Th1 and Th2 subpopulations. Patients with lymphatic filariasis express a spectrum of diseases that range from an asymptomatic state of microfilaremia in which the larval stage of the parasite circulates in high numbers in the blood stream and thus allows transmission of disease by the mosquito vector, to the symptomatic condition of chronic lymphatic obstruction of tropical pulmonary eosinophilia [90]. In this latter group of individuals the disease process of chronic lymphatic obstruction (CP) or tropical pulmonary eosinophilia is associated with marked parasite antigen-driven T-cell responses and elevated serum levels of parasite antigen-specific immunoglobulin levels. This contrasts to the antigen-specific immune *hypo*responsiveness observed among individuals with the asymptomatic microfilaremia (MF) [91, 92]. Prior studies have postulated the presence of parasite-derived suppressive factors or a subpopulation of suppressor cells; these studies, however, have not been well substantiated in subsequent studies [91, 93–95]. Recently it has been proposed that individuals with MF have developed a state of immune tolerance based on the observation of decreased frequencies of antigen-reactive T cells using a limiting dilution analysis [96]. However, the examination of the relative frequencies of antigen-driven Th2 and Th1 CD4+ subsets in these two clinical manifestations of lymphatic filariasis provides a more compelling framework to understand the nature of immune response and potentially in other helminth infections. As shown in table 4, MF patients have markedly lower frequencies of antigen-specific CD4+ cells capable of producing IFN-γ (e.g. a Th1 response), a finding that would account for the impaired antigen-driven T-cell proliferation [92, 96, 97] and IFN-γ and IL-2 production previously seen [91]. Of particular interest is the similar frequencies of IL-4 producing CD4+ cells among MF and CP patients indicating a shift toward a predominantly Th2 response. This shift is supported by the observation that MF patients have markedly elevated serum levels of polyclonal IgE and IgG4 antibodies (generally associated with increased Th2 responses). The biological significance of this shift may be important on several accounts. First, the down-regulation of a Th1 response may diminish induction of cellular effectors that contribute to the pathology of the disease and facilitate removal of the microfilaria. This might, in part, be accomplished by the production of irrelevant IgE that saturates the mast cells' FcεR and renders it unable to be triggered by parasite antigen. Alternatively, the high serum levels of IgE may be directed to a very restricted subset of filarial antigens (perhaps MF excretory/secretory or surface antigens) that may be in such low concentrations in crude parasite antigen preparations of adult worms used for detection of parasite-specific IgE responses that these levels are missed in our

Table 4. Frequency of parasite antigen-induced cytokine secreting lymphocytes in patients with different clinical manifestations of lymphatic filariasis

Pathology	n	Frequency of antigen-induced cytokine-secreting lymphocytes[1]	
		IL-4	IFN-γ
Microfilaremia	6	1/4,770 (1/38,462–1/1,366)	1/8,955 (1/41,667–1/3,257)
Chronic lymphatic obstruction	10	1/10,614 (1/71,429–1/2,203)	1/1,525[2] (1/7,194–1/457)

[1]Expressed as geometric means.
[2]Significantly different from microfilaremics, p = 0.03.

current assays. Furthermore, these responses may cross-react with the infective larvae surface antigens and thus induce a state of concomitant immunity mediated by IgE, but directed primarily toward infective larvae (L3). Although there is no current information with regard to differential or preferential recognition of L3 antigens by MF, this notion of concomitant immunity is consistent with observations made in both human studies [98–100] and in some animal studies (e.g. *Trichinella spiralis* in rats [101]).

The mechanism for this shift to a Th2 response may result from not only an expansion of Th2-reactive clones, but also from the down-regulation of Th1 responses. An important cross-regulatory candidate molecule is IL-10, a cytokine produced by Th2 cells (as well as mast cells and Ly1+ B cells in mice) that can act to inhibit cytokine synthesis by Th1 cells and CD8+ lymphocytes [10] perhaps by a mechanism of diminished antigen presenting capacity of monocytes through down-regulation of major histocompatibility complex class II (MHC II [102]). It has been demonstrated that CD4+ cells from schistosome-infected mice produce large quantities of IL-10 and that the addition of neutralizing anti-IL-10 mAb to antigen-stimulated cell cultures from infected mice caused a dramatic augmentation in IFN-γ synthesis [61]. These observations in the murine system appear to parallel those observed in our studies on human filarial infection. In individuals infected with *W. bancrofti*, the addition of neutralizing anti-human IL-10 mAb restored the ability of PBMC to proliferate to parasite antigens in many of the hyporesponsive MF individuals; there was no corresponding effect on proliferation responses by PBMC from CP patients [unpubl. observations].

Schistosomiasis

The pattern of Th subset responses in experimental infections of schistosomiasis in murine models has been described above [59, 103]. Although the cytokine responses in human schistosomiasis have been less well studied, existing evidence suggests a similar pattern to that observed in murine schistosome models. In the study of an individual with acute schistosomiasis, a marked induction of parasite antigen-driven T-lymphocyte response with increased IL-4 and IFN-γ production was observed (table 3). With chronic infections, antigen-specific T-cell proliferation is depressed with impaired parasite antigen-driven IFN-γ and IL-2 production and augmented IL-4 production to mitogens (parasite antigen-driven IL-4 production was undetectable) in culture supernatants of PBMC [104, 105]. The magnitude of the IL-4 response was correlated with both the intensity of infection and serum IgE levels, and was inversely related to mitogen-driven IFN-γ production [105]. Three to 6 months after parasitologic cure with antischistosomal therapy, parasite antigen-driven T-cell proliferation [106] as well as IFN-γ and IL-2 production in PBMC culture supernatants increased significantly [104] indicating a depression of these responses during chronic infections. No change in mitogen-driven IL-4 production was observed after chemotherapy, however [105]. Interestingly, a group of individuals infected with either schistosomiasis or intestinal nematodes whose treatment failed had an increased ratio of mitogen-driven IL-4 to IFN-γ production in culture supernatants compared to individuals successfully cured [105]. This modulation of the immune response through suppression of Th1 responses can be interpreted as a host mechanism to reduce the pathology associated with the inflammatory response to infection. If this suppression is great enough, particularly in heavy infections, complete cure may not be achieved because cell-mediated responses controlled by Th1 cells may not be sufficiently engaged to eliminate the parasite. Alternatively, patients with heavy infections, and the highest ratio of IL-4 to IFN-γ, may simply not eliminate all the parasites with a single treatment. It is tempting to interpret these observations in human schistosomiasis with those in the murine models; that the Th2 response generated with chronic infection depresses pathology, but is not important in immunity, and a Th1 response induces cell responses associated with eliminating the parasite and protection. However, studies of *Schistosoma haematobium* in the Gambia showed that individuals with high serum levels of parasite-specific IgE after antischistosomal chemotherapy were less likely to become reinfected or have lighter reinfections compared to individuals with lower IgE levels [107]. Additionally, it was observed that there was an associated increase in serum levels of parasitic-specific IgE with increasing age, which was closely correlated with development of acquired resistance. This study suggests that an augmented Th2 response, indicated by elevated serum IgE levels may, in fact, be associated with the development of protective immunity. An important deficiency in the Gambian study, however, was the failure to assess the magnitude of bladder pathology, the site of egg elimination in *S. haematobium* infections, because a heavy infection can potentially lead to substantial scarring and shunting of blood through the hemorrhoidal plexus resulting in an apparent decline in egg output. Thus, the high serum IgE levels may simply reflect heavier worm burdens.

Clearly, the examination of human helminth infections in the context of cytokine production by Th subsets and its relationship to pathology and immunity remains poorly understood. As the study of lymphatic filariasis provides a good system to investigate the role of CD4+ cells in modulating the immune response and pathology, *S. haematobium* infection provides probably the best system for studying the biologic role of Th responses in the development of protective immunity because of its distinct epidemiologic profile of infection, the fact that infected individuals can be readily cured of infection with one course of therapy, the easy accessibility of urine samples to determine egg burdens and the sensitivity of ultrasound to detect anatomic abnormalities in the urinary tract.

Mechanisms Involved in Determining Th Subset Differentiation

The Role of Antigen Structure

The finding that T cells from the same individual when stimulated with an allergen or a parasite antigen expand Th2-like CD4+ cells and when stimulated with a different type of soluble antigen (e.g., TT or PPD) expand Th1-like cells suggest the structural properties of a given antigen may be the determinant that causes the preferential activation of a particular T-cell subset. Indeed, in murine models of *S. mansoni* infection, Th2 responses appear to be induced by the schistosome eggs whereas Th1 responses are directed at the schistosomula [59]. In a murine model of leishmaniasis, two distinct fractions of soluble crude leishmania extracts have been demonstrated to induce preferentially either Th1- or Th2-like responses when used to immunize mice prior to *Leishmania* infection [reviewed in 108–110]. In other studies of *Leishmania*, an octamer constructed of 10-amino acid long peptides corresponding to a tandemly repeated region of a *L. major* glycoprotein stimulated a Th2 response when injected subcutaneously into sensitized mice [111]. This contrasted with induction of a Th1 response to a specific leish manial surface protein indicating that certain epitopes can preferentially induce a subset of Th cells [112]. The authors, however, were careful to point out that these results may be biased due to the mode of antigen presentation, because the octamer, for example, generated the Th2 response only when administered subcutaneously and evoked little or no response when given intravenously.

Although a number of allergens and parasite antigens have been cloned and sequenced, the epitopes responsible for the induction of Th1 or Th2 responses have not been sought in any systematic fashion. Nevertheless, their preferential recognition by IgE suggests an association with a Th2-like response. However, a comparison of their primary structures has failed to reveal a common structural motif to suggest that a particular peptide sequence might be associated with induction of a Th2 response [113]. Recently, rutin, a polyphenol-containing compound, was shown to induce IgE through the production of IL-4 (and not IL-2) [114] suggesting that structures other than peptides can contribute to the physicocochemical character of particular antigens that may influence the ultimate nature of the immune response it engenders. A study of different epitopes of filarial paramyosin identified IgE binding domains that in some cases differed from those that bound IgG [115]. Whether those epitopes that preferentially induced IgE (in contrast to those that induced IgG) were responsible for activating T cells to make Th2 types of cytokines awaits further study.

Antigen Presentation

Differences in antigen processing have also been used to explain why some antigens induce Th2 cytokines while other antigens cause activation of Th1 cells. In murine systems, Th2 T-cell clones appear to respond to antigen presented by B cells better than Th1 T-cell clones although presentation by macrophages was equivalent between the two CD4$^+$ subsets [116]. Conversely, hepatic cells enriched for Kupffer cells present antigen more efficiently to Th1 cells [117]. Similarly, brain capillary endothelial cells are more effective at antigen presentation to Th2 cells whereas brain capillary smooth muscle cells present antigen better to Th1 cells [118]. Additional studies have implicated Langerhans' cells as particularly good antigen presenters to Th2 clones [119]. Although none of these studies has been in humans, a similar phenomenon might be expected to occur.

Not only does one have to consider the nature of the antigen-presenting cells in the antigen presentation process, but also one must consider the role of chronic antigenic stimulation, a phenomenon that is found commonly in both atopic diseases and parasitic infection. Parasites, for example, such as invasive helminths, characteristically produce chronic infections and as such have adopted mechanisms that do not induce protective immune responses. Similarly, in allergic diseases the immune response is such that with repeated natural antigenic exposure the effect is a gradual but steady deterioration rather than improvement. One hypothesis to explain these observations has been to implicate chronic low dose antigenic stimuli as the cause. Chronic local or systemic stimulation with antigens, it is said, might alter the manner in which these antigens are processed and presented to the immune system such that Th2-type responses are induced preferentially. Although this issue has not as yet been approached experimentally, it remains an important fundamental theory for explaining the types of responses seen in atopy and parasitic infection.

Conclusions

The observation that two distinct and cross-regulatory CD4$^+$ T-cell subsets exist based on patterns of cytokines they produce has provided a powerful and intuitively appealing model to explain the classical immuno-logical divergence between the development of humoral immune responses and that of cell-mediated immune responses. The frequently observed reciprocal relationships between antibody production (or Th2 functions) and cell-mediated immunity (or Th1 responses) found in many infectious disease systems can be explained by the cross-regulatory actions of certain cytokines, most notably IL-10, TGF-β, and IL-4 that have been shown to enhance antibody formation and suppress classical cellular immune responses and IFN-γ that can mediate the converse. This dichotomy has caught the attention of those studying tropical diseases, in that high levels of peripheral blood eosinophils and serum IgE production, often associated with depressed cellular immune responses, are characteristics of chronic helminth infec-tions. Conversely, in some individuals the same helminth infections can induce partial or total protective immunity, whether natural or vaccine-induced, that are associated with Th1-like responses. A 'successful' helminth infection generates a strong Th2 response – a response that enables the parasite to persist in the host without causing itself harm, a strategy particularly suited to helminths since they do not replicate within the host. In contrast, in the absence of cell-mediated Th1-like response, bacterial or protozoan infections that can replicate within the host would produce overwhelming infection and death, as has been observed in some experimental models and occasionally in humans.

A critical question is what determines if a helminth infection produces immunity or persistence of infection, particularly if successful vaccines are to be produced. A range of factors have been postulated including: (1) early exposure to the parasite (e.g. tolerance); (2) genetic makeup of the host; (3) the route of infection and subsequent migration of larval stages resulting in unusual antigen presentation mechanisms that strongly favor Th2 responses; (4) unique characteristics of particular helminth antigens, and (5) the persistence of antigenic stimulus of the parasite leading to selective polarization of the T-cell subsets. Studies to examine these factors are in the early phases and have broader implications as to how to alter the balance between the Th1 and Th2 responses so that protective immunity could be enhanced or that atopic and autoimmune diseases could be treated.

Finally, a particularly intriguing question remains as to the functional significance of the marked IgE and eosinophilia observed in chronic helminth infections. If as postulated above, the generation of a Th2 response is to modulate cellular responses to enable the persistence of the parasite and reduce pathology in the host, then the immediate hypersensitivity responses may either participate in this immunomodulatory role or act to inhibit further infection by infective stages of the parasite (e.g. concomitant immunity). Clearly, delineation of the functional role of the Th2 subset and the immediate hypersensitivity responses it engenders may provide insight as to how outcome of specific infectious challenges to the host immune defenses is determined.

References

1 Banchereau J, Defrance T, Galizzi JP, Miossec P, Rousset F: Human interleukin-4. Bull Cancer (Paris) 1991;78:299–306.
2 Pene J, Rousset F, Briere F, Chretien I, Bonnefoy J, Spits H, Yokota T, Arai N, Arai K, Banchereau J, de Vries JE: IgE production by normal human lymphocytes is induced by interleukin-4 and suppressed by interferons γ and α and prostaglandin E₂. Proc Natl Acad Sci USA 1988;85:6880–6884.
3 Vercelli D, Geha RS: Regulation of IgE synthesis in humans. J Clin Immunol 1989; 9:75–83.

4 Clutterbuck E, Hirst E, Sanderson C: Human interleukin-5 (IL-5) regulates the production of eosinophils in human bone marrow cultures: comparison and interaction with IL-1, IL-3, IL-6, and GM-CSF. Blood 1989;73:1504–1512.

5 Sanderson CJ: The biological role of interleukin-5. Int J Cell Cloning 1990;1:147–153.

6 Mosmann TR, Cherwinski H, Bond MW, Giedlin MA, Coffman RL: Two types of murine helper T cell clone I. Definition according to profiles of lymphokine activities and secreted proteins. J Immunol 1986;136:2348–2357.

7 Mosmann TR, Coffman RL: TH1 and TH2 cells: different patterns of lymphokine secretion lead to different functional properties. Annu Rev Immunol 1989;7:145–173.

8 Street NE, Mosmann TR: Functional diversity of T lymphocytes due to secretion of different cytokine patterns. FASEB J 1991;5:171–177.

9 Firestein GS, Roeder WD, Laxer JA, Townsend KS, Weaver CT, Hom JT, Linton J, Torbett BE, Glasebrook AL: A new murine CD4+ cell subset with an unrestricted cytokine profile. J Immunol 1989;143:518–525.

10 Mosmann TR, Moore KW: The role of IL-10 in cross-regulation of TH1 and TH2 responses. Immunol Today 1991;12:A49–A53.

11 Finkelman FD, Pearce EJ, Urban JJ, Sher A: Regulation and biological function of helminth-induced cytokine responses. Immunol Today 1991;12:A62–A66.

12 Vercelli D, Leung DY, Jabara HH, Geha RS: Interleukin-4-dependent induction of IgE synthesis and CD23 expression by the supernatants of a human helper T cell clone. Int Arch Allergy Appl Immunol 1989;88:119–121.

13 Romagnani S, Maggi E, Del PG, Parronchi P, Macchia D, Tiri A, Ricci M: Role of interleukin-4 and gamma-interferon in the regulation of human IgE synthesis: possible alterations in atopic patients. Int Arch Allergy Appl Immunol 1989;88:111–113.

14 Nutman TB, Volkman DJ, Hussain R, Fauci AS, Ottesen EA: Filarial parasite-specific T cell lines: induction of IgE synthesis. J Immunol 1985;134:1178–1184.

15 Maggi E, Del PG, Tiri A, Macchia D, Parronchi P, Ricci M, Romagnani S: T cell clones providing helper function for IgE synthesis release soluble factor(s) that induce IgE production in human B cells: possible role for interleukin-4 (IL-4). Clin Exp Immunol 1988;73:57–62.

16 Maggi E, Del PG, Macchia D, Parronchi P, Tiri A, Chretien I, Ricci M, Romagnani S: Profiles of lymphokine activities and helper function for IgE in human T cell clones. Eur J Immunol 1988;18:1045–1050.

17 Romagnani S: Regulation and deregulation of human IgE synthesis. Immunol Today 1990;11:316–321.

18 King CL, Ottesen EA, Nutman TB: Cytokine regulation of antigen-driven immunoglobulin production in filarial parasite infections in humans. J Clin Invest 1990; 85:1810–1815.

19 O'Hehir RE, Bal V, Quint D, Moqbel R, Kay AB, Zanders ED, Lamb JR: An in vitro model of allergen-dependent IgE synthesis by human B lymphocytes: comparison of the response of an atopic and a non-atopic individual to Dermatophagoides spp. (house dust mite). Immunology 1989;66:499–504.

20 Mahanty S, Abrams JS, King CL, Limaye AP, Nutman TB: Parallel regulation of IL-4 and IL-5 in human helminth infections. J. Immunol, in press.

21 King CL, Gallin JI, Malech HL, Abramson SL, Nutman TB: Regulation of immuno

globulin production in hyperimmunoglobulin E recurrent-infection syndrome by interferon gamma. Proc Natl Acad Sci USA 1989;86:10085–10089.

22 Souillet G, Rousset F, de Vries JE: Alpha-interferon treatment of patient with hyper-IgE syndrome. Lancet 1989;i:1384.

23 Coffman RL, Seymour BWP, Hudak S, Jackson J, Rennick D: Antibody to interleukin-5 inhibits helminth-induced eosinophilia in mice. Science 1989;245:308–310.

24 Finkelman FD, Holmes J, Katona IM, Urban JJ, Beckmann MP, Park LS, Schooley KA, Coffman RL, Mosmann TR, Paul WE: Lymphokine control of in vivo immuno-globulin isotype selection. Annu Rev Immunol 1990;8:303–333.

25 Sher A, Coffman RL, Hieny S, Cheever AW: Ablation of eosinophil and IgE responses with anti-IL-5 or anti-IL-4 antibody fails to affect immunity against *Schistosoma mansoni* in the mouse. J Immunol 1990;145:3911–3916.

26 Sher A, Coffman RL, Hieny S, Scott P, Cheever AW: Interleukin-5 (IL-5) is required for blood and tissue eosinophilia but not granuloma formation induced by infection with *Schistosoma mansoni*. Proc Natl Acad Sci USA 1990;87:61–66.

27 Korenaga M, Hitoshi Y, Yamaguchi N, Sato Y, Takatsu K, Tada I: The role of interleukin-5 in protective immunity to *Strongyloides venezuelensis* infection in mice. Immunology 1991;72:502–507.

28 Limaye AP, Abrams JS, Silver JE, Ottesen EA, Nutman TB: Regulation of parasite-induced eosinophilia: selectively increased interleukin-5 production in helminth-infected patients. J Exp Med 1990;172:399–402.

29 Groopman JE: Status of colony-stimulating factors in cancer and AIDS. Semin Oncol 1990;17:31–37.

30 Denburg JA, Silver JE, Abrams JS: Interleukin-5 is a human basophilopoietin: induction of histamine content and basophilic differentiation of HL-60 cells and peripheral blood basophil-eosinophil progenitors. Blood 1991;77:1462–1467.

31 Saito H, Hatake K, Dvorak AM, Leiferman KM, Donnenberg AD, Arai K, Ishizaka K, Ishizaka T: Selective differentiation and proliferation of hematopoietic cells induced by recombinant human interleukins. Proc Natl Acad Sci USA 1988; 85:2288–2292.

32 Limaye AP, Abrams JS, Silver JE, Awadzi K, Francis HF, Ottesen EA, Nutman TB: Interleukin-5 and the posttreatment eosinophilia in patients with onchocerciasis. J Clin Invest 1991;88:1418–1421.

33 Mazza G, Thorne KJI, Richardson BA, Butterworth AE: The presence of eosinophil-activating mediators in sera from individuals with *Schistosoma mansoni* infections. Eur J Immunol 1991;21:901–905.

34 Owen WF, Rothenberg ME, Petersen J, Weller PF, Silberstein D, Sheffer AL, Stevens RL, Soberman RJ, Austen KF: Interleukin-5 and phenotypically altered eosinophils in the blood of patients with the idiopathic hypereosinophilic syndrome. J Exp Med 1989;170:343–348.

35 Owen WJ, Petersen J, Sheff DM, Folkerth RD, Anderson RJ, Corson JM, Sheffer AL, Austen KF: Hypodense eosinophils and interleukin-5 activity in the blood of patients with the eosinophilia-myalgia syndrome. Proc Natl Acad Sci USA 1990;87:8647–8651.

36 Zielinski RM, Lawrence WD: Interferon-alpha for the hypereosinophilic syndrome. Ann Intern Med 1990;114:338–339.

37 Murphy PT, Fennelly DF, Stuart M, O'Donnell JR: Alpha-interferon in a case of hypereosinophilic syndrome. Br J Haematol 1990;75:619–620.

38 Jarrett EE, Miller HR: Production and activities of IgE in helminth infection. Prog Allergy. Basel, Karger, 1982, vol 31, pp 178–233.

39 Askenase PW: Immune inflammatory responses to parasites: the role of basophils, mast cells and vasoactive amines. Am J Trop Med Hyg 1977;26:96–107.

40 Askenase PW: Immunopathology of parasitic diseases: involvement of mast cells and basophils. Springer Semin Immunopathol 1980;2:417–442.

41 Kay AB: The role of the eosinophil. J Allergy Clin Immunol 1979;64:90–104.

42 Kay AB: Eosinophils: role in asthma, allergy and parasite immunity. N Engl Reg Allergy Proc 1985;6:341–345.

43 Moqbel R: Helminth-induced intestinal inflammation. Trans R Soc Trop Med Hyg 1986;80:719–727.

44 Moqbel R, Sass KS, Goetzl EJ, Kay AB: Enhancement of neutrophil- and eosinophil-mediated complement-dependent killing of schistosomula of Schistosoma mansoni in vitro by leukotriene B_4. Clin Exp Immunol 1983;52:519–527.

45 Anwar AR, McKean JR, Smithers SR, Kay AB: Human eosinophil- and neutrophil-mediated killing of schistosomula of Schistosoma mansoni in vitro. I. Enhancement of complement-dependent damage by mast cell-derived mediators and formyl methionyl peptides. J Immunol 1980;124:1122–1129.

46 Beaven MA: Histamine: its role in physiological and pathological processes. Monogr Allergy. Basel, Karger, 1978, vol 13, pp 1–113.

47 Moqbel R, King SJ, MacDonald AJ, Miller HR, Cromwell O, Shaw RJ, Kay AB: Enteral and systemic release of leukotrienes during anaphylaxis of Nippostrongylus brasiliensis-primed rats. J Immunol 1986;137:296–301.

48 Miller HR, Huntley JF, Wallace GR: Immune exclusion and mucus trapping during the rapid expulsion of Nippostrongylus brasiliensis from primed rats. Immunology 1981;44:419–429.

49 Miller HR, Huntley JF: Protection against nematodes by intestinal mucus. Adv Exp Med Biol 1982;144:243–245.

50 Miller HR: The protective mucosal response against gastrointestinal nematodes in ruminants and laboratory animals. Vet Immunol Immunopathol 1984;6:167–259.

51 Miller HR: Gastrointestinal mucus, a medium for survival and for elimination of parasitic nematodes and protozoa. Parasitology 1987;94:577–600.

52 Castro GA: Immunological regulation of epithelial function. Am J Physiol 1982;243:321–329.

53 Ishizaka T, Saito H, Hatake K, Dvorak AM, Leiferman KM, Arai N, Ishizaka K: Preferential differentiation of inflammatory cells by recombinant human interleukins. Int Arch Allergy Appl Immunol 1989;88:46–49.

54 Kirshenbaum AS, Goff JP, Dreskin SC, Irani AM, Schwartz LB, Metcalfe DD: IL-3-dependent growth of basophil-like cells and mast-like cells from human bone marrow. J Immunol 1989;142:2424–2429.

55 Madden KB, Urban JFJ, Zilterner HJ, Schrader JW, Finkelman FD, Katona IM: Antibodies to IL-3 and IL-4 suppress helminth-induced intestinal mastocytosis. J Immunol 1991;147:1387–1391.

56 Tsai M, Takeishi T, Geissler N, Langley KE, Zsebo KM, Galli SJ: Stem cell factor, a ligand for c-kit, promotes mast cell proliferation and maturation in vitro and in vivo. FASEB J 1991;5:A1086.

57 Thompson-Snipes L, Dhar V, Bond MW, Mosmann TR, Moore KW, Rennick D: Interleukin-10: a novel stimulatory factor for mast cells and their progenitors. J Exp Med 1991;173:507.

58 Hultner L, Druez C, Moeller J, Uyttenhove C, Schmitt E, Rude P, Dormer P, van Snick J: Mast cell growth-enhancing activity is structurally related and functionally identical to the novel mouse T cell growth factor P40/TCGFIII (interleukin-9). Eur J Immunol 1990;20:1413.

59 Pearce EJ, Caspar P, Grzych JM, Lewis FA, Sher A: Down-regulation of Th1 cytokine production accompanies induction of Th2 responses by a parasitic helminth, *Schistosoma mansoni*. J Exp Med 1991;173:159–166.

60 Grzych JM, Pearce EJ, Cheeve A, Caulada Z, Caspar P, Hieny S, Lewis S, Sher A: Egg deposition is the major stimulus for the production of Th2 cytokines in murine *Schistosoma mansoni*. J Immunol 1991;146:1322.

61 Sher A, Fiorentino D, Casper P, Pearce E, Mosmann T: Production of IL-10 by CD4$^+$ lymphocytes correlates with down-regulation of Th1 cytokine synthesis in helminth infection. J Immunol 1991;147:2713–2716.

62 Henderson GS, Lu X, McCurley TL, Colley DG: In vivo molecular analysis of lymphokines involved in murine response during *Schistosoma mansoni* infection II. Quantification of IL-4 mRNA, IFN-γ mRNA, and IL-2 mRNA levels in granulomatous livers, mesenteric lymph nodes, and spleens during the course of modulation. J Immunol 1992;148:2261–2269.

63 Domingo EO, Warren KS: Endogenous desensitization: changing host granulomatous response to schistosome eggs at different stages of infection with *Schistosoma mansoni*. Am J Pathol 1968;52:369–377.

64 Capron A, Dessaint JP, Capron M, Ouma JH, Butterworth AE: Immunity to schistosomes: progress toward vaccine. Science 1987;238:1065–1072.

65 Mahmoud AA, Warren KS, Peters P: A role for the eosinophil in acquired resistance to *Schistosoma mansoni* infection as determined by anti-eosinophil serum. J Exp Med 1975;142:805.

66 Othman MI, Higashi GI: In vivo role of eosinophils in *Schistosoma mansoni* immunity in chronic murine infections. American Society of Tropical Medicine and Hygiene Annual Meeting, New Orleans, 1990.

67 Urban JFJ, Katona IM, Paul WM, Findelman RD: Interleukin-4 is important in protective immunity to a gastrointestinal nematode infection in mice. Proc Natl Acad Sci USA 1991;88:5513–5517.

68 O'Hehir RE, Lamb JR: MHC class II and allergen-specific T cell clones. Clin Exp Allergy 1991;1:173–177.

69 Wierenga EA, Snoek M, Bos JD, Jansen HM, Kapsenberg ML: Comparison of diversity and function of house dust mite-specific T lymphocyte clones from atopic and non-atopic donors. Eur J Immunol 1990;20:1519–1526.

70 Nutman TB, Ottesen EA, Fauci AS, Volkman DJ: Parasite antigen-specific human T cell lines and clones. Major histocompatibility complex restriction and B cell helper function. J Clin Invest 1984;73:1754–1762.

71 Del Prete G, Tiri A, Maggi E, De CM, Macchia D, Parronchi P, Rossi ME, Pietrogrande MC, Ricci M, Romagnani S: Defective in vitro production of gamma-interferon and tumor necrosis factor-alpha by circulating T cells from patients with the hyperimmunoglobulin E syndrome. J Clin Invest 1989;84:1830–1835.

72 Quint DJ, Bolton EJ, McNamee LA, Solari R, Hissey PH, Champion BR, MacKenzie AR, Zanders ED: Functional and phenotypic analysis of human T-cell clones which stimulate IgE production in vitro. Immunology 1989;67:68–74.

73 Raghavachar A, Fleischer S, Frickhofen N, Heimpel H, Fleischer B: T lymphocyte

control of human eosinophilic granulopoiesis. Clonal analysis in an idiopathic hypereosinophilic syndrome. J Immunol 1987;139:3753–3758.

74 Maggi E, Macchia D, Parronchi P, Del PG, De CM, Piccinni MP, Simonelli C, Biswas P, Romagnani S, Ricci M: The IgE response in atopy and infections. Clin Exp Allergy 1991;1:72–78.

75 Maggi E, Del PG, Parronchi P, Tiri A, Macchia D, Biswas P, Simonelli C, Ricci M, Romagnani S: Role for T cells, IL-2 and IL-6 in the IL-4-dependent in vitro human IgE synthesis. Immunology 1989;68:300–306.

76 Maggi E, Mazzetti M, Ravina A, Simonelli C, Parronchi P, Macchia D, Biswas P, Di PM, Romagnani S: Increased production of IgE protein and IgE antibodies specific for fungal antigens in patients with the acquired immunodeficiency syndrome. Ric Clin Lab 1989;19:45–49.

77 Maggi E, Biswas P, Del PG, Parronchi P, Macchia D, Simonelli C, Emmi L, De CM, Tiri A, Ricci M, et al: Accumulation of Th-2-like helper T cells in the conjunctiva of patients with vernal conjunctivitis. J Immunol 1991;146:1169–1174.

78 Tepper RI, Levinson DA, Stanger BZ, Campos TJ, Abbas AK, Leder P: IL-4 induces allergic-like inflammatory disease and alters T cell development in transgenic mice. Cell 1990;62:457–467.

79 Del Prete G, Tiri A, Mariotti S, Parronchi P, Pinchera A, Romagnani S, Ricci M: Thyroiditis as a model of organ-specific autoimmune disease. Clin Exp Rheumatol 1989;4:267–276.

80 Romagnani S, Maggi E, Del Prete G, Parronchi P, Macchia D, Tiri A, Ricci M: Role of interleukin-4 and gamma-interferon in the regulation of human IgE synthesis: possible alterations in atopic patients. Int Arch Allergy Appl Immunol 1989;88:111–113.

81 Nutman TB, Hussain R, Ottesen EA: IgE production in vitro by peripheral blood mononuclear cells of patients with parasitic helminth infections. Clin Exp Immunol 1984;58:174–182.

82 Wierenga EA, Snoek M, De Groot C, Chretien I, Bos JD, Jansen HM, Kapsenberg MS: Evidence for compartmentalization of functional subsets of CD4$^+$ lymphocytes in atopic patients. J Immunol 1990;144:4651–4656.

83 Parronchi P, Macchia D, Piccinni M, Biswas P, Simonelli C, Maggi E, Ricci M, Ansari AA, Romagnani S: Allergen- and bacterial antigen-specific T-cell clones established from atopic donors show a different profile of cytokine production. Proc Natl Acad Sci USA 1991;88:4538–4542.

84 Del Prete G, De Carli M, Mastromauro C, Biagiotti R, Macchia D, Falagiani P, Ricci M, Romagnani S: Purified protein derivative of *Mycobacterium tuberculosis* and excretory-secretory antigen(s) of *Toxocara canis* expand in vitro human T cells with stable and opposite (type 1 T helper or type 2 T helper) profile of cytokine production. J Clin Invest 1991;88:346–350.

85 Haanen JBAG, de Waal Malefijt R, Res PCM, Kraakman EM, Ottenhoff THM, de Vries RRP, Spits H: Selection of a human T helper type 1-like cell subset by mycobacteria. J Exp Med 1991;174:583–592.

86 Yssel H, Shanafelt M, Soderberg C, Schneider PV, Anzola J, Peltz G: *Borrelia burgdorferi* activates a T helper type 1-like T cell subset in lyme arthritis. J Exp Med 1991;174:593–601.

87 Bradley LM, Duncan DD, Tonkonogy S, Swain SL: Characterization of antigen-specific CD4$^+$ effector T cells in vivo: immunization results in a transient population

of MEL-14-, CD45RB-helper cells that secretes interleukin-2 (IL-2), IL-3, IL-4, and interferon-γ. J Exp Med 1991;174:547–559.

88 Swain SL, Weinberg AD, English M: CD4⁺ T cell subsets. Lymphokine secretion of memory cells and of effector cells that develop from precursors in vitro. J Immunol 1990;144:1788–1799.

89 King CL, Thyphronitis G, Nutman TB: Enumeration of IgE secreting B cells. A filter spot-ELISA. J Immunol Methods 1990;132:37–43.

90 Ottesen EA: Immunopathology of lymphatic filariasis in man. Springer Semin Immunopathol 1980;2:373–385.

91 Nutman TB, Kumaraswami V, Ottesen EA: Parasite-specific anergy in human filariasis. Insights after analysis of parasite antigen-driven lymphokine production. J Clin Invest 1987;79:1516–1523.

92 Piessens WF, McGreevy PB, Piessens PW, McGreevy M, Koiman I, Saroso HS, Dennis DT: Immune responses in human infections with *Brugia malayi*. Specific cellular unresponsiveness to filarial antigens. J Clin Invest 1980;65:172–179.

93 Piessens WF, Ratiwaytano S, Tuti S, Palmieri JH, Piessens PW, Koiman I, Dennis DT: Antigen-specific suppressor cells and suppressor factors in human filariasis with *Brugia malayi*. N Engl J Med 1980;302:833–837.

94 Piessens WF, Partono F, Hoffman SL, Ratiwayanto S, Piessen PW, Palmieri JR, Koiman I, Dennis DT, Carney WP: Antigen-specific suppressor T lymphocytes in human lymphatic filariasis. N Engl J Med 1982;307:144–148.

95 Nutman TB, Kumaraswami V, Pao L, Narayanan PR, Ottesen EA: An analysis of in vitro B cell immune responsiveness in human lymphatic filariasis. J Immunol 1987;138:3954–3959.

96 King CL, Kumaraswami V, Poindexter RW, Kumari S, Jayaraman K, Alling DW, Ottesen EA, Nutman TB: Immunologic tolerance in lymphatic filariasis: Diminished parasite-specific T and B lymphocytes precursor frequency in the microfilaremic state. J Clin Invest 1992;89:(in press).

97 Ottesen EA, Weller PF, Heck L: Specific cellular immune unresponsiveness in human filariasis. 1977;33:413–421.

98 Bundy DAP, Grenfell BT, Rajagopalan PK: Immunoepidemiology of lymphatic filariasis: the relationship between infection and disease. Immunol Today 1991;12:A71–A75.

99 Bradley DJ, McCullough FS: Egg output and epidemiology of *Schistosoma haematobium*. II. An analysis of the epidemiology of endemic *S. haematobium*. Trans R Soc Trop Med Hyg 1973;67:491–500.

100 Smithers SR, Terry RJ: Resistance to experimental infection with *Schistosoma mansoni* in rhesus monkeys induced by the transfer of adult worms. Trans R Soc Trop Med Hyg 1967;61:517–523.

101 Despommier D: Immunity to *Trichinella spiralis*. Am J Trop Med Hyg 1977;26:68.

102 de Waal Malefyt R, Haanen J, Spits H, Roncarolo M, te Velde A, Figdor C, Johnson K, Kastelein R, Yssel H, de Vries J: Interleukin-10 (IL-10) and viral IL-10 strongly reduce antigen-specific human T cell proliferation by diminishing the antigen-resenting capacity of monocytes via down-regulation of class II major histocompatibility complex expression. J Exp Med 1991;174:915–924.

103 James SL, Sher A: Cell-mediated immune response to schistosomiasis. Curr Top Microbiol Immunol 1990;155:21.

104 Zwingenberger K, Irschick E, Siqueira Vergetti JG, Correia Decal AR, Janssen-Rosseck R, Bienzle U, Huber C, Feldmeier H: Release of interleukin-2 and gamma-interferon by peripheral mononuclear cells in human *Schistosomiasis mansoni* infection normalizes after chemotherapy. Scand J Immunol 1989;30:463–471.

105 Zwingenberger K, Hohmann A, Cardoso de Brito M, Ritter M: Impaired balance of interleukin-4 and interferon-γ production in infections with *Schistosoma mansoni* and intestinal nematodes. Scand J Immunol 1991;34:243–251.

106 Colley DG, Barsoum IS, Dahawi HSS, Gamil F, Habib M, El Alamy MA: Immune responses and immunoregulation in relation to human schistosomiasis in Egypt. III. Immunity and longitudinal studies of in vitro responsiveness after treatment. Trans R Soc Trop Med Hyg 1986;80:952–957.

107 Hagan P, Blumenthal UJ, Dunn D, Simpson AJG, Wilkins HA: Human IgE, IgG4 and resistance to reinfection with *Schistosoma haematobium*. Nature 1991;349:243–245.

108 Scott P, Pearce E, Cheever AW, Coffman RL, Sher A: Role of cytokines and CD4$^+$ T-cell subsets in the regulation of parasite immunity and disease. Immunol Rev 1989;112:161–182.

109 Locksley RM, Scott P: Helper T-cell subsets in mouse leishmaniasis: induction, expansion and effector function. Immunol Today 1991;12:A58–61.

110 Locksley RM, Heinzel FP, Holaday BJ, Mutha SS, Reiner SL, Sadick MD: Induction of Th1 and Th2 CD4$^+$ subsets during murine *Leishmania major* infection. Res Immunol 1991;142:28–32.

111 Liew FY, Millott SM, Schmidt JA: A repetitive peptide of *Leishmania* can activate T helper type 2 cells and enhance disease progression. J Exp Med 1990;172:1359–1365.

112 Yang DM, Fairweather N, Button LL, McMaster WR, Kahl LP, Liew FY: Oral *Salmonella typhimurium* (AroA–) vaccine expressing a major leishmanial surface protein (gp63) preferentially induces T helper 1 cells and protective immunity against leishmaniasis. J Immunol 1990;145:2281–2285.

113 Francus T, Siskind GW, Becker CG: Role of antigen structure in the regulation of IgE isotype expression. Proc Natl Acad Sci USA 1983;80:3430.

114 Baum CG, Szabo P, Siskind GW, Becker CG, Firpo A, Clarick CJ, Francus T: Cellular control of IgE induction by a polyphenol-rich compound. Preferential activation of Th2 cells. J Immunol 1990;145:779–784.

115 Steel C, Limberger RJ, McReynolds LA, Ottesen EA, Nutman TB: B cell responses to paramyosin. Isotypic analysis and epitope mapping of filarial paramyosin in patients with onchocerciasis. J Immunol 1990;145:3917–3923.

116 Chang TL, Shea CM, Urioste S, Thompson RC, Boom WH, Abbas AK: Heterogeneity of helper/inducer T lymphocytes. III. Responses of IL-2 and IL-4-producing (Th1 and Th2) clones to antigens presented by different accessory cells. J Immunol 1990;145:2803–2808.

117 Magilavy DB, Fitch FW, Gajewski TF: Murine hepatic accessory cells support the proliferation of Th1 but not Th2 helper T lymphocyte clones. J Exp Med 1989;170:985–990.

118 Fabry Z, Sandor M, Gayewsky TF, Lynch R, Hart M: CD4$^+$ T cell subsets are differentially activated when the antigen is presented by murine brain capillary endothelium versus smooth muscle cells. FASEB J 1990;4:A1802.

119 Daynes RA, Araneo BA, Dowell TA, Huang K, Dudley D: Regulation of murine lymphokine production in vivo. III. The lymphoid tissue microenvironment exerts regulatory influences over T helper cell function. J Exp Med 1990;171:979–996.

120 Dessein AJ, Parker WL, James SL, David JR: IgE antibody and resistance to infection. I. Selective suppression of the IgE antibody response in rats diminishes the resistance and the eosinophil response to *Trichinella spiralis* infection. J Exp Med 1981;153:423–436.

Christopher L. King, MD, PhD, Division of Geographic Medicine, Case Western Reserve University, 2109 Adelbert Road, Cleveland, OH 44106 (USA)

Coffman RL (ed): Regulation and Functional Significance of T-Cell Subsets.
Chem Immunol. Basel, Karger, 1992, vol 54, pp 166–211

Delayed-Type Hypersensitivity Recruitment of T Cell Subsets via Antigen-Specific Non-IgE Factors or IgE Antibodies: Relevance to Asthma, Autoimmunity and Immune Responses to Tumors and Parasites[1]

Philip W. Askenase[2]

Section of Allergy and Clinical Immunology, Department of Medicine,
Yale University School of Medicine, New Haven, Conn., USA

Introduction

Delayed-type hypersensitivity (DTH) is an important in vivo manifestation of T cell-mediated immunity. In humans and mice these responses are principally due to CD4+ T cells [1–3]. In DTH responses, CD4+ T cells act extravascularly to recruit circulating bone marrow-derived nonspecific effector leukocytes at a site of antigen challenge. Depending upon circumstances, the recruited effector leukocytes can be monocyte/macrophages, neutrophils, eosinophils, basophils or mixtures. DTH mechanisms by which CD4+ T cells recruit effector leukocytes are central to deleterious inflammation that characterizes autoimmunity and allergy, as well as protective immune recruitment of effector leukocytes to local sites of invasion by microbes, parasites or tumors. A dramatic example of the importance of such protective in vivo CD4+ T cell-dependent responses is in AIDS patients who lack such responses and therefore suffer from a variety of opportunistic infections and neoplasias that eventually lead to death.

Our laboratory has had a particular research focus in DTH responses. We began with the fact that CD4+ T cells were known to recruit effector

[1] Supported in part by grants from the NIH (AI-12211 and AI-26689).

[2] The author is most grateful to Gregory Geba, Robert Mann, Sylvette Nazario and Scott Roberts for reviewing the manuscript and to Marilyn Avallone for her superb secretarial skills.

leukocytes into extravascular tissues. Normally there are few T cells in the tissues. Furthermore, no mechanism exists for the recirculation of the entire T cell repertoire to the extravascular peripheral nonlymphoid tissues. However, very few CD4$^+$ T cells need to be recruited extravascularly in order to subsequently recruit the other cells that constitute greater than 99% of the infiltrate. In fact, one study employing limiting dilution analysis of local passive transfer of T cell clones with antigen has shown that very few CD4$^+$ DTH-effector T cells (i.e. one T cell) are sufficient for eliciting DTH [4]. Thus, we asked: How do the CD4$^+$ T cells themselves get recruited extravascularly? We have found that a new cell, which we call a DTH-initiating cell, governs the recruitment of CD4$^+$ T cells. The mechanisms for DTH initiation are just beginning to be understood. Thus far we have identified two pathways by which DTH initiation can occur. In the first, these DTH-initiating cells act by releasing an antigen-binding factor that functions *like* IgE antibody in its ability to sensitize cutaneous mast cells for the release of serotonin, a vasoactive amine. Released serotonin alters vessels in the local microenvironment to allow recruitment of CD4$^+$ T cells. In the other mechanism, IgE antibody itself can, under some circumstances, initiate DTH.

The DTH Cascade

We have likened DTH responses to a cascade of cellular and molecular events (fig. 1). By a cascade we mean a process that begins with a trickle and ends with a torrent. The DTH cascade can really be divided into at least two sequential components: the DTH-initiating phase and the DTH-effector phase. The late-occurring effector phase is the classical, CD4$^+$ T cell-mediated, delayed component of DTH. In the systems we have studied, specifically sensitized CD4$^+$ T cells are induced within the lymph nodes and/ or spleen by day 4 after cutaneous, subcutaneous or intravenous immunization and then pass into the circulation. These are classical, antigen/MHC class II-restricted CD3$^+$, CD4$^+$ T cells expressing αβ T cell receptors (TCR) [3]. In the periphery, at the site of a DTH response, these αβ T cells must be recruited out of the vessels into the extravascular tissues. They then interact with local antigen-presenting cells (APC) (such as Langerhans' cells in the skin), via their αβ TCR. The αβ TCR bind to complexes of processed antigen peptides and MHC class II molecules on the surface of APC. This results in the activation of these specific T cells and the release of a characteristic

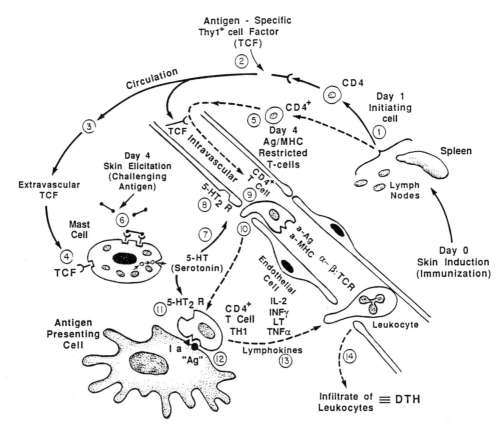

Fig. 1. Effector cascade for elicitation of DTH responses in the skin of mice. Circled numbers indicate the postulated sequence of individual steps. Solid arrows indicate the sequence of early, DTH-initiating steps. Dotted arrows indicate the sequence of steps mediated by late-acting, classical DTH-effector T cells.

profile of nonspecific cytokines. In cutaneous DTH in mice, these T cells produce a Th-1 cytokine profile of: IL-2, interferon-γ, TNFβ (lymphotoxin) [2], and probably TNFα. These cytokines and others then lead to the local recruitment of circulating, nonspecific, effector leukocytes into the extravascular tissues. The recruited leukocytes are derived from the bone marrow and constitute the classical perivascular infiltrate of nonspecific inflammatory cells that characterizes DTH.

Our work suggests that all of this classical late component of DTH, due to CD4+ T cells, is necessary but not sufficient for elicitation of DTH. In addition, *prior* local release of serotonin is needed [5–10]. Serotonin has two loci of action. The first is on the vessels to open gaps, vasodilate and perhaps increase expression of adhesion molecules [5, 7, 11]. The second is on serotonin-2 receptors on the surface of the recruited Th-1 T cells [10]. Serotonin may act as a co-stimulus, or second signal, that along with signal one provided by cross-linking of αβ TCR, allows for lymphokine transcription and production. Thus, DTH can be elicited if sensitized, polyclonal, Th-1 T cells are isolated from DTH-initiating cells, and are injected into the local extravascular space, along with antigen, by local *passive* transfer. *Importantly*, no DTH is elicited if these same cells are injected intravenously and the skin is challenged [3]. However, if to this systemic intravenous adoptive transfer of isolated late-acting αβ T cells is added, either: DTH-initiating cells [3, 12]; the soluble, antigen-specific, non-IgE DTH-initiating factors they produce in vitro [13]; or small amounts of IgE antibody [14]; or if the local site is challenged simultaneously by injecting small amounts of serotonin [14], then these intravascular, CD4+, Th-1 T cells are able to be recruited locally to elicit DTH.

These various components, that need to be added to the adoptive transfers of αβ T cells, provide what we call DTH initiation, and constitute the early cascade of DTH. This is mediated usually by DTH-initiating cells that are induced *within 1 day* of immunization in the lymph nodes and spleen [15–17]. In fact, in all systems in which this has been studied, DTH-initiating cells are induced *within 1 day* [12, 15–21]. These DTH-initiating cells are antigen-specific, Thy-1+, but CD4-, and are surface-negative for many other T cell markers [3, 13]. DTH-initiating cells do not need to recirculate, but are induced in lymph node and spleen where they *release* an antigen-specific DTH-initiating factor that has yet to be characterized molecularly, but has biological activity that resembles IgE antibody [9, 11, 21–27]. This factor is released into the circulation [26] where it can be detected and distinguished from antibody, such as IgE. Our evidence indicates that the DTH-initiating factor then passes into the extravascular space to sensitize cutaneous mast cells via surface-binding sites for a postulated constant region of the DTH-initiating factor [7, 9, 11, 23, 25, 27]. Thus, when the skin is challenged by antigen it is thought that antigen binds to these IgE-like DTH-initiating factors on the surface of mast cells. This causes a specialized form of mast cell activation that leads to differential release of serotonin, in preference to histamine, without much degranulation [9, 11]. Serotonin appears to be released via cytosolic vesicles [11]. Work on model systems in vitro, and electron microscopy of mast cells in DTH, has led us to hypothesize that serotonin is transported via vesicles out of the granules by specific serotonin-binding proteins, and is released subsequently by vesicular exocytosis [11, 28–33]. Released serotonin is the final mediator of DTH initiation. Local serotonin for DTH initiation can also be provided via specific IgE antibody and antigen [14], or via direct injection of serotonin [14].

In the new work to be summarized below, five particular aspects of DTH initiation will be reviewed: (1) the phenotype of DTH-initiating cells; (2) description of DTH-initiating cell clones; (3) details concerning DTH initiation by monoclonal IgE antibody; (4) preliminary evidence that mast cells, as well as platelets, can be sources of DTH-initiating serotonin, and (5) the relevance of DTH initiation to asthma, autoimmunity and immune responses to tumors and parasites.

The Phenotype of DTH-Initiating Cells

The system in which we have examined DTH initiation most thoroughly is contact sensitivity in mice. In this system, mice are sensitized by topical application of a high concentration of a contact sensitizer such as picryl chloride (5%, PCl) or oxazolone (3%, OX) applied to the abdomen, chest and paws. These contact sensitizers are reactive organic haptens that bind to free amino groups on host carrier proteins creating neoantigens, and also induce local nonspecific inflammation. Generally, 4 days after contact painting, actively sensitized animals are challenged on the ears with a dilute solution (0.8%) of PCl or OX in olive oil, and ear swelling responses are measured subsequently at 24 h with an engineer's micrometer. Figure 2 shows the in vivo adoptive cell transfer assay that was used to determine the phenotype of DTH-initiating cells. We harvested contact sensitivity effector cells (lymph node and spleen cells), from actively sensitized animals at day 4. Approximately 5×10^7 cells were transferred intravenously (IV) to recipients that were subsequently challenged on the ears with 0.8% PCl. Challenge generally was 1 day later to allow DTH-initiating cells to produce DTH-initiating factor to sensitize mast cells in the skin [15–17]. With this procedure, the time course of DTH initiation in recipients was similar to the donors. There was an onset of ear swelling within 30 min, that peaked at 2 h and, in many cases, diminished over 4–6 h. Subsequently, a second late or delayed wave of ear swelling due to recruited αβ T cells, peaked at 18–24 h after challenge and was still present at 48 and even 72 h. The early, DTH-initiating component was due to antigen-specific, Thy-1[+] cells and was also mast cell and serotonin-

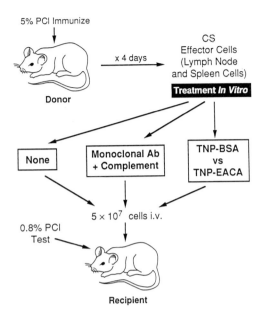

5% PCI Immunize

x 4 days

CS
Effector Cells
(Lymph Node
and Spleen Cells)

Treatment In Vitro

Donor

None

Monoclonal Ab
+ Complement

TNP-BSA
vs
TNP-EACA

5×10^7 cells i.v.

0.8% PCI
Test

Recipient

Fig. 2. Assay for determining the phenotype of DTH-initiating versus DTH-effector T cells via in vitro treatment of transferring cells with various mAb and complement, or for determining the hapten versus hapten/carrier specificity of DTH-initiating versus DTH-effector T cells via desensitization of transferring cells by treatment in vitro with hapten-conjugated amino acid versus protein.

dependent [15–17]. From these findings the DTH cascade hypothesis presented above was formulated.

To determine the phenotype of the DTH-initiating cells, transferring cells were treated in vitro with various monoclonal antibodies (mAb) and then separated either by complement cytotoxicity, panning, or anti-immunoglobulin beads (fig. 2). Antibodies that depleted only the late component thus defined epitopes on DTH-effector T cells. Antibodies that depleted transfer of both components were found to react with determinants on *both* DTH-initiating and late-acting DTH-effector T cells. *However, the use of mAb that only reacted with determinants on DTH-initiating cells by causing depletion of DTH-initiating cells, led to an inability to elicit late DTH.* Using this technique, we determined that the phenotype of the two Thy-1⁺ cells that act sequentially to mediate DTH was different. The early DTH-initiating cell, that produced 2 h edematous ear swelling responses, had a most unusual phenotype for an antigen-specific cell of: Thy-1^+, CD5^+, CD4^-, CD8^-, CD3^-, TCR$\alpha\beta^-$, TCR$\gamma\delta^-$, B220^+ (CD45RA$^+$), IL-2R$^-$, and IL-3R$^+$ [3, 17]. In contrast, the late-acting, classical, DTH-effector T cell, that provided optimal responses at 24 h, had the expected phenotype of: Thy-1^+, CD5^+, CD4^+, CD8^-, CD3^+, TCR$\alpha\beta^+$, TCR$\gamma\delta^-$, B220^-, IL-2R$^+$, and IL-3R$^-$ [3, 17].

It is important to point out that in the latter experiments, which employed anti-B220 (CD45RA) or anti-IL-3R, and found that the DTH-initiating cell was positive, and the late DTH-effector T cell was negative, that *both* components were absent in recipients because of a lack of DTH initiation. To demonstrate that treatment with these latter antibodies and complement was affecting *only* the early-acting, DTH-initiating cell, these treated populations were transferred together with a population of *1-day immune cells* that was *only* able to transfer DTH-initiating activity. This *reconstituted* the ability of the late-acting effector T cells, that remained after treatment with anti-B220 or anti-IL-3R antibodies, to transfer DTH [3, 12, 13].

Antigen Receptors on DTH-Initiating and DTH-Effector T Cells

Our results indicate that the antigen receptor on the DTH-initiating cells and the DTH-effector T cells are different. The late cell seems to have a hapten/carrier-specific receptor that is undoubtedly the $\alpha\beta$ TCR, whereas the DTH-initiating cell has antigen receptors that are hapten-specific. In vitro antigen-specific desensitization has suggested a hapten-specific receptor on DTH-initiating cells. In these experiments, 4-day immune, contact-sensitized effector cells were treated in vitro with various conjugates of hapten-amino acid, or hapten-protein, at 100 µg/ml for 60 min at 37 °C [34]. The cells were then washed and transferred intravenously. The recipient ears were, in turn, challenged immediately (fig. 2). With this protocol, the early component occurred at about 6–8 h (the time to produce DTH-initiating factors and sensitize the periphery) [15–17], and the late component was measured at 24 h.

Using this system, we found that 4-day PCl (trinitrophenyl (TNP) chloride) immune cells that were desensitized in vitro with either TNP-bovine serum albumin (TNP-BSA; hapten-protein conjugate) or TNP-ϵ amino caproic acid (TNP-EACA; hapten-amino acid conjugate) were unable to transfer *either* the early or late components of DTH. In contrast, desensitization in vitro with dinitrophenyl-BSA (DNP-BSA) or DNP-lysine allowed full expression of the early and late components [35]. A cell mixing and transfer system enabled us to determine whether early or late cells were affected by in vitro desensitization with specific antigens. We found that TNP-BSA desensitization affected the early *and* late cells in DTH. Thus, when 4-day PCl immune cells were treated in vitro with TNP-BSA, recipients had neither early nor late reactivity. When these desensitized cells were reconstituted with 1-day immune cells, *only* early reactivity was elicited, as a consequence of the added 1-day immune cells. No late reactivity was brought out. In contrast, desensitization using TNP-EACA (a hapten-amino acid conjugate) was found to affect the early cell, and *not* the late cell. In these experiments, 4-day PCl immune cells that were desensitized in vitro with TNP-EACA failed to produce early or late reactivity in recipients. However, addition of 1-day PCl immune cells to the TNP-EACA desensitized 4-day PCl immune cells provided early reactivity that *reconstituted* expression of late 24-hour DTH. This was due to the fact that late 24-hour DTH-effector cell activity remained in the 4-day PCl immune cells that had been desensitized in vitro with TNP-EACA. Finally, when the 1-day PCl immune cells were desensitized in vitro with TNP-EACA, they no longer could reconstitute DTH activity in 4-day PCl immune cells that had been treated in vitro with TNP-EACA [35].

In conclusion, the in vitro antigen desensitization experiments suggest that there is a hapten/carrier receptor, probably αβ TCR, on late-acting DTH-effector T cells. For this desensitization, hapten-carrier antigen is probably taken up in vitro by APC in the spleen and lymph node population that then desensitize the late-acting DTH-effector T cells in vitro, or preclude their ability to respond rapidly to antigen again *after* transfer and skin challenge. In contrast, hapten-amino acid conjugates cannot interact effectively with αβ TCR on these cells, but are able to interact with a presumably hapten-specific receptor that is present on the surface of DTH-initiating cells, and that may be related to the antigen (hapten)-specific factor that is released by these cells, and which initiates DTH.

Study of DTH Initiation in Nude Mice and in
Severe Combined Immunodeficiency Mice

Important insights into DTH-initiating cells was provided by studies in immunodeficient mice. We found that antigen-specific DTH-initiating cells could be induced in some strains of athymic nude mice [36]. PCl contact sensitization in these nude mice resulted in the ability to elicit antigen-specific early 2-hour ear swelling after local PCl challenge; i.e. similarly PCl immunized nude mice that were topically challenged with OX did not respond. In contrast, euthymic littermate nu/+ mice immunized with PCl and challenged with PCl elicited comparable early responses and *also* large late 24-hour responses. These late responses undoubtedly were due to αβ T cells which were severely depleted in their nu/nu counterparts. Correspondingly, nu/nu mice that were contact-sensitized with OX just elicited early OX-specific ear swelling and no response to PCl.

In contrast, immunization of severe combined immunodeficiency (SCID) mice, which lack all B cells and T cells [37, 38], did not induce early or late reactivity, compared to BALB/c (Baily strain) controls [36] or CB17 congenic controls [Geba and Askenase, in preparation]. Hence, our studies of DTH initiation in immunodeficient mice showed that athymic nude mice could be induced to express antigen-specific 2-hour cutaneous responses. Moreover, an antigen-specific DTH-initiating factor could be derived from the lymphoid organs of sensitized nude mice [36]. In contrast, nude mice could not elicit the late 24-hour component of DTH. Furthermore, the early *and* late components of DTH were absent in SCID mice. We concluded that DTH-initiating cells were relatively *thymic-independent* since they could be induced in nude mice. However, since these cells could not be induced in SCID mice, which have a defect in joining the coding regions of rearranged gene segments [37–39], these studies suggested that rearranging genes were responsible for the expression of antigen specificity by DTH-initiating cells.

Deriving DTH-Initiating Cell Clones

In the cloning of DTH-initiating cells, we took advantage of several characteristics of contact sensitivity responses in nude mice. Firstly, DTH-initiating cells occurred in the absence of large numbers of αβ T cells [36]. Secondly, in contrast to normal mice in which DTH-initiating cells are rapidly inhibited by CD8$^+$ suppressor T cells, these suppressor T cells did not exist in nude mice [40]. Thus, DTH-initiating cells were not suppressed, and in fact could be *boosted* in nude mice. We therefore sensitized and boosted nude mice by weekly contact painting with OX. Lymphoid cells were then harvested and expanded in vitro with IL-3 (WEHI-3 supernatant) since we knew that DTH-initiating cells had IL-3R [3]. A line was created by selecting among these cells for Thy-1$^+$, CD4$^-$, CD8$^-$ cells, and immortalizing by adding

Moloney murine leukemia virus. Finally, clones were isolated from one of the lines by employing a FACS sorter that delivered a single cell that was doubly positive for both Thy-1 and B220 [13]. This eliminated from the cloning any classical T cells (Thy-1$^+$, B220$^-$) or classical B cells (Thy-1$^-$, B220$^+$) that remained in the line.

About 50 well-growing clones were developed and then screened by FACS for phenotype. Among these, 12 were selected for testing by the in vivo assay of DTH initiation. Two clones were found to be capable of DTH initiation. Most of our subsequent studies have employed one of these clones that we call WP-3.27 [13]. In our initial experiments with this clone we again employed the cell transfer of 4-day contact sensitivity effector cells that in this case were OX-sensitized. When such cells were treated with complement alone they transferred both early and late reactivity. When they were treated in vitro with anti-B220 and complement they transferred neither early nor late responses due to depletion of DTH-initiating cells. The late DTH activity of cells remaining could be restored by adding 1-day OX-immune cells, or, *importantly*, the clone WP-3.27 [13]. In contrast, controls that received anti-B220-treated, late DTH-acting, isolated αβ T cells, that were combined with WP-3.27 cloned cells, and were challenged at the same time on the ears *with PCl*, failed to elicit early, or late reactivity. In a subsequent study, we showed that DTH initiation could be mediated by early-acting cells that were antigen mismatched (or MHC incompatible) with late-acting DTH-effector T cells. Moreover, we showed that the clone could initiate DTH for PCl-specific late-acting DTH-effector T cells if the recipients were challenged with *both* OX and PCl, but not when they were challenged with OX *or* PCl alone [12]. Therefore, we concluded that clone WP-3.27 systemically transferred antigen-specific 2-hour reactivity that allowed expression of specific late 24-hour DTH activity.

In further experiments, supernatant derived from the clone WP-3.27 was antigen affinity purified. We initially cleared the supernatants by passage through a TNP-antigen column, and then applied the filtrate to an OX-antigen column, that was thoroughly washed and then eluted with base. The resulting OX-F derived from WP-3.27 was added to isolated late-acting DTH-effector T cells remaining after anti-B220/plus complement treatment. This reconstituted DTH in recipients challenged with OX, but not in recipients challenged with PCl [13]. We concluded that the antigen-binding OX-F derived from clone 3.27 systemically transferred antigen-specific 2-hour reactivity that allowed expression of specific, late 24-hour DTH.

The antigen-specific DTH-initiating clones were then phenotyped by FACS. These studies largely confirmed previous data that had been obtained by mAb depletion of the ability of polyclonal cells to transfer ear swelling activity. Thus, we found that the antigen-specific DTH-initiating clone had the following phenotype: (a) for T-like markers: Thy-1$^+$, CD5$^+$, CD4$^-$, CD8$^-$, CD3$^-$, TCRαβ$^-$, TCRγδ$^-$ [13, 41]; (b) for B-like markers: B220$^+$ (CD45RA$^+$), and sIg$^-$; (c) for growth factor receptors: IL-2R$^-$, and IL-3R$^+$; (d) there were also macrophage-like markers (class II$^+$, FcγR$^+$, MAC I$^+$), that apparently were acquired by Moloney leukemia virus transformation of the clone [41]. We concluded that the DTH-initiating cell had a most unusual phenotype for an antigen-specific cell and that the phenotype suggested a primitive thymic-independent cell.

Molecular Studies on DTH-Initiating Clones

The molecular phenotype of the clones was characterized by Northern blot analysis using cDNA probes that encode chains of antigen-specific molecules, and by Southern analysis of DNA rearrangements [13, and Rhamabhadran and Askenase, in preparation]. As a control, we showed that the DTH-initiating clones transcribed mRNA for Thy-1 and IL-3 receptors, but not for CD3γ, δ, ε, or ζ. Transcripts for κ and λ immunoglobulin light chains were not found, nor was there mRNA for IgCμ. We were particularly interested in IgCε which also was negative by Northern analysis. For TCR chains, the clone did not transcribe TCR-Cα, Cβ, nor Cγ mRNA. However, there were transcripts for components of TCR δ, namely Cδ and Jδ1, but not Jδ2. However, these transcripts were of low abundance and abnormally large size (3 kb with respect to normal 1.8-kb transcripts for known γδ T cell clones). Furthermore, Southern analysis, employing the Jδ1 probe that provided a positive result on Northern analysis, failed to reveal rearrangement of the δ locus in the clones. Finally, the TCR γ locus, the IgH locus, and the κ and λ light chain loci were not rearranged. We concluded that the large δ-mRNA transcripts were germ line transcripts that were spliced.

In summary, we are puzzled as to which rearranging genes might encode the antigen-specific OX factor produced by clone WP-3.27. We think that rearranging genes are involved because the factors are induced by immunization with specific antigen, act antigen specifically and are antigen-binding [22–26]. In addition, analogous factors can be induced in normal and nude mice, but not in SCID mice [36]. However, immunoglobulin genes do not seem to be involved as the clone is surface Ig-negative and does not transcribe nor rearrange immunoglobulin heavy or light chain genes. Likewise, TCR $\alpha\beta$ does not seem to be involved since the clone is negative for $C\alpha$ or $C\beta$ mRNA, as well as surface and mRNA-negative for components of CD3. Finally, a conventional $\gamma\delta$ T cell receptor does not seem to be involved because the cell is surface CD3-negative. Moreover, although δ mRNA is transcribed, the δ locus and the γ locus are not rearranged. Thus, if rearranging genes encode the antigen-specific DTH-initiating factor, they are presumably novel and are the subject of current investigations in our laboratory.

Our conclusions about DTH-initiating cells are: (1) they are important and required for elicitation of cellular immunity in vivo; as exemplified by DTH; (2) they appear to be primitive cells since they are of a mixed phenotype, and in particular are IL-3R$^+$, which is a property of pre-T and pre-B cells, but not mature T or B cells; (3) they are relatively thymic-independent as they can be induced in athymic nude mice. We hypothesized that these may be primitive pre-T cells, since the genes that encode the antigen-specific DTH-initiating factor are presumed to be rearranged, but at the moment are unknown. The reasons for postulating that rearranged genes encode the DTH-initiating factors are: (1) they are antigen-specific and antigen-binding; (2) they are inducible by antigen-specific immunization, and (3) they are made by euthymic and athymic mice, but not by SCID mice that have a defect in rearranging gene segments. Of course it is possible that the antigen specificity we have determined by the procedures we have used depends on mechanisms other than conventional DNA rearrangements. These important points cannot be settled until these genes are isolated and sequenced.

Is the DTH-Initiating Factor IgE Antibody?

The preceding experiments established that DTH initiation is due to non-IgE antigen-specific factors that are produced by DTH-initiating, non-B, non-T cells. Because of the biologic similarity of DTH-initiating factors to IgE antibodies, we inquired whether actual IgE antibodies themselves could initiate DTH. That is, could IgE initiate T cell DTH; at least in some instances? Actually, our original hypothesis about the DTH-initiating factor and the DTH-initiating cells was that the factor was IgE antibody and that DTH-initiating cells were IgE-producing plasma cells. In the early 1980s we discovered an early immediate hypersensitivity-like aspect of DTH responses, and showed that this early component of DTH was dependent on serotonin and mast cells, and was due to an antigen-specific factor that was produced by lymphoid cells. Thus, our principal hypothesis at that time was that this was due to IgE antibody produced by B cells [22–27].

Thus, we questioned whether DTH was an example of pure T cell-mediated immunity and quite separate from humoral immunity that is transferable by soluble serum antibodies. We noted that all of the older arguments about the absence of a role of antibodies in DTH responses were really arguments about IgG antibodies. We postulated that a small amount of IgE antibody, produced by a small number of IgE-producing B cells, could 'sneak through' all of the previous arguments against antibody and B cells playing a role in DTH. Specifically, although the transfer of DTH was certainly due to T cells, it was not considered that there might be a few contaminating *T cell-dependent* B cells, that could make a small amount of IgE antibody, upon which T cell-mediated DTH was dependent. Thus, the presence of DTH in animals treated with B cell-depleting high doses of cyclophosphamide (200 mg/kg) [42, 43], or in animals rendered agammaglobulinemic by treatment from birth with anti-μ chain antisera (so-called μ-suppressed mice) [44], might have been due to IgE-forming B cells, which are relatively resistant to cyclophosphamide [45, 46], and more difficult to inhibit via μ suppression than IgG responses [47].

Thus, we set out to determine if the DTH-initiating factor was IgE antibody. Employing the techniques we could use at that time, our studies showed that IgE antibody was not present. Namely: (1) IgCε determinants could not be detected in our factor by an ELISA assay, nor by a sensitive radioimmunoassay [22, 23]. (2) Like IgE, the DTH-initiating factor was inactivated by heating to 56 °C, but was not sensitive to reduction and alkylation, which is uniquely true of IgE antibodies [23–26]. (3) The DTH-initiating factor, passed through anti-immunoglobulin and anti-IgE columns, was retained by and could be eluted from anti-factor antibody columns that did not retain true IgE antibodies, either monoclonal or polyclonal [23, 27]. (4) IgE antibody activity was still evident on cutaneous mast cells 48 h after transfer. Alternatively, DTH-initiating factor activity was only evident for a few hours after transfer, and was gone by 48 h [23]. Finally, (5) the factor mediated differential release of serotonin versus histamine, while IgE caused equal release of both mediators [9]. Thus, in these early studies, we concluded that the DTH-initiating factor was not IgE antibody.

Subsequently, our work with polyclonal DTH-initiating cells showed the phenotype of these cells was not consistent with a classical B cell [3]. Furthermore, the DTH-initiating clones we have described recently had a phenotype that did not resemble a B cell [13]. They did not have surface Ig, and did not transcribe nor rearrange Ig heavy or light chain genes. However, we recently reopened this question when it was shown that IgE antibodies can sometimes be produced without the DNA rearrangements usually needed for switch recombination [48, 49], and that there are CD5[+] B cells [50], and even Thy-1[+] B cells [51].

Employing a new set of techniques appropriate to the current era, we again have concluded that the DTH-initiating factor is not IgE and that DTH-initiating cells are not B cells. Namely, we have developed a sensitive quantitative ELISA for TNP-specific IgE (sensitivity 50 pg/ml) that has failed to find a rise in TNP-specific IgE in the serum of PCl-sensitized mice, nor in the supernatants from sensitized cells harvested 4 days following contact sensitization [Gardner and Askenase, in preparation]. There is no IgE detectable in our factor preparations by this ELISA, nor by Western blotting with monoclonal anti-IgE that can detect as little as 2 ng of IgE. Finally, we have performed Northern analysis on the DTH-initiating clone for mRNA transcripts of IgE constant region genes, and this analysis has been negative. Recently we have expanded this to polymerase chain reaction (PCR). We have found that this very sensitive technique failed to reveal any mRNA that could generate cDNA that upon PCR amplification could hybridize with a cDNA probe encoding IgE heavy chain [Ramabhadran and Askenase, in preparation]. Furthermore, PCR of 4-day immune lymphoid cells revealed only a very small amount of IgE mRNA that did not increase with contact sensitization. Thus, we have concluded once again that the DTH-initiating factor is not IgE, and that the DTH-initiating cell is not an IgE-producing B cell.

DTH Initiation by Monoclonal IgE Antibody

In newer studies we have examined the issue of IgE in another way by asking whether actual IgE antibodies could act like the DTH-initiating factor and therefore initiate DTH. The protocol we have used to test whether IgE can initiate DTH is based on our ability to deplete DTH-initiating cells with anti-B220 mAb and complement. In this experiment, 4-day PCl immune (TNP-chloride immune), late-acting DTH-effector T cells were depleted of

Group	I.V. Transfer of PCl-Immune Cells Treated with anti-B220+C	I.V. Injection of TNP-Specific IgE Antibody into Cell Recipients (dose / mouse)	Contact Sensitivity Response in Adoptive Recipients (Challenge with PCl) (units x 10⁻³ cm ± SE)
A	+	none	
B	+	0.1 ng	
C	+	1 ng	
D	+	10 ng	
E	–	100 ng	
F	+	100 ng	
G	–	10 µg	
H	+	10 µg	
J	–	100 µg	
I	+	100 µg	

Fig. 3. Mediation of DTH initiation by purified TNP-specific monoclonal IgE antibody; the effect of IgE dose. Four-day PCl-immune cells were treated with monoclonal anti-B220 antibody plus complement to provide isolated late-acting DTH T cells. These cells were transferred alone into mice (group A), or into mice that were injected 1 day previously with different doses of a purified preparation of monoclonal IgE antibody (hatched bars). Then recipients were topically skin challenged with PCl. Each dose of IgE (except 0.1–10 ng) was also transferred alone to separate groups of mice that were challenged similarly with PCl (open bars). A negative control group was just challenged with PCl. Resultant background ear swelling at 2 and 24 h in this group was subtracted from responses in the other groups to provide the net ear swelling that is shown in the figure. $p < 0.01$, group C 24 h versus group A 24 h; $p < 0.0005$, group D 24 h versus group A 24 h.

DTH-initiating cells by treatment in vitro with anti-B220 mAb and complement. This isolated population of polyclonal late-acting DTH-effector T cells was transferred IV to mice that also received various doses of monoclonal anti-TNP IgE antibody. Following this transfer of TNP-specific IgE, plus PCl-immune late-only T cells, the ears of recipient mice were challenged with PCl [14].

We found that IgE antibody did and did not initiate DTH (fig. 3). Examination of recipients of different doses of IgE antibody alone showed that the minimal amount of systemically given IgE that elicited a detectable early 2-hour ear swelling response was about 0.1 µg. Progressive 10-fold higher doses sensitized for progressively higher immediate responsiveness, such that 100 µg gave a very large (maximal) early ear swelling response in which the ears obviously were quite edematous. When these 10-fold dilutions of IgE were combined with late-acting DTH effector cells that remained after treatment with anti-B220 and complement, there was a very interesting dose response. At 0.1–1 µg IgE per mouse there were small early responses due to IgE that were followed by definite large 24-hour responses which were quite significant compared to controls that received cells treated with anti-B220 and complement, but no IgE [14]. *Thus, IgE could initiate DTH, and the optimal dose of IgE seemed to be about 1 µg of IgE per mouse.*

Experiments were done to examine the lower dose limits. It was found that doses of IgE as small as 10 or even 1 ng/mouse could initiate DTH. Interestingly, these lower doses of IgE were not associated with any detectable early *macroscopic* ear swelling, Thus, DTH initiation can take place in the absence of any detectable early macroscropic reaction [14]. We have been fortunate to be able to detect *its macroscopic manifestations* in the contact hypersensitivity systems we have been employing in mice.

There is another interesting part to the dose-response curve. IgE doses of 10 or 100 µg/recipient, that mediated very large local immediate reactions, were associated with greatly *diminished* late DTH responses. We also found a similar dose-response curve with monoclonal IgG1/anti-TNP antibody [14], the other isotype of mouse immunoglobulins for which there are Fc receptors on mast cells, and are capable of activating mast cells for release of mediators such as serotonin. Thus, DTH initiation was mediated at 0.1 or 1 µg of anti-TNP IgG1/per mouse, and was inhibited by larger doses, namely 10 and 100 µg of specific IgG1 mAb per mouse [14].

Respective Positive and Negative Roles of Serotonin and Histamine in DTH Initiation

We formulated a hypothesis to explain the pathogenesis of the failure of high doses of IgE or IgG1 to initiate DTH. Our past studies indicated that mast cell-derived serotonin is an important final mediator of DTH initiation [5–11]. We also showed previously that the DTH-initiating factors led to a preferential release of serotonin versus histamine from mast cells; a phenomenon we have called 'differential release' [9, 28–33]. In contrast, it is well known that IgE causes compound sequential exocytosis of mast cell granules and therefore an equal percent release of *both* histamine and serotonin [9]. Furthermore, we considered that histamine is well known to inhibit many T cell functions, including DTH in mice, probably by acting on inhibitory histamine-2 receptors on the surface of T cells and leading to inhibition of cytokine gene transcription and production through elevation of cyclic AMP [53–57]. Thus, we hypothesized that the higher doses of IgE (or IgG1) led not only to serotonin release, but also to local release of *large amounts* of histamine that acted to inhibit recruited DTH-effector T cells, accounting for the inhibition of DTH when high doses of IgE (or IgG1) were used to initiate these responses.

The protocol we used to test if high doses of IgE failed to initiate DTH because of histamine-2 suppression was as follows: We transferred monoclonal anti-TNP IgE antibody IV at a high dose (100 μg/mouse), versus an optimal dose (1 μg/mouse), together with isolated late-acting DTH T cells that remained after treatment with anti-B220 mAb and complement. Recipients of this combination of an optimal versus a high dose of IgE, together with late DTH T cells, were treated with subcutaneous injections of cimetidine, a histamine-2 receptor antagonist, or saline [14]. It was postulated that cimetidine might block the inhibitory effects of high doses of histamine that could be released by the higher doses of IgE antibody. We then challenged the recipients with PCl. We found (fig. 4) that an optimal dose of IgE (1 μg/mouse) led to optimal initiation of DTH, that was not influenced by treating recipients with cimetidine. On the other hand, transfer of 100 μg of IgE, along with late-acting T cells, produced very poor late 24-hour responses that were *fully restored* to the level seen with an optimal dose of IgE, if recipients of the high dose of IgE *and* late-acting T cells *also* were treated with cimetidine [14].

Thus, a high dose of IgE, leading to a large release of *both* serotonin and histamine by sequential exocytosis, can initiate DTH when there is inhibition of histamine-2 receptors, that are presumed to be on the surface of the late-acting DTH effector T cells, and lead to inhibition of cytokine production. Cimetidine blockade of histamine-2 receptors on the recruited αβ T cells allows for an effect that is like differential release in that only the vasoactivity of serotonin occurs, since the capillaries of mice are insensitive to histamine.

These experiments led us to investigate whether local injection of *serotonin alone* could initiate DTH. Thus, we transferred IV isolated late-acting DTH T cells remaining after treatment of 4-day immune cells with anti-B220 and complement to recipients that simultaneously were chal-

Group	Transfer of PCl-Immune Cells Treated with Anti-B220+C	Injection of TNP-Specific IgE Antibody Into Cell Recipients (μg/mouse)	Treatment with Cimetidine (50mg/kgx2)	24 Hr Contact Sensitivity Response In Adoptive Recipients (Challenge with PCl) (units x 10⁻³ cm ± SE)
A	+	None	–	
B	+	None	+	
C	–	1	–	
D	–	1	+	
E	–	100	–	
F	–	100	+	
G	+	1	–	
H	+	1	+	
I	+	100	–	
J	+	100	+	

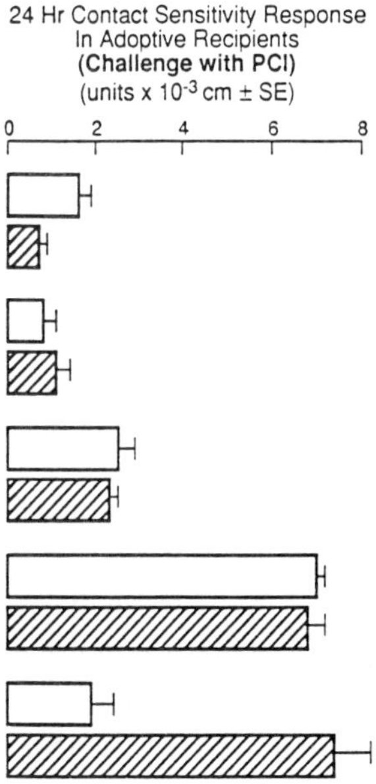

Fig. 4. Treatment with the histamine-2 receptor antagonist cimetidine restores late DTH in mice with DTH initiation mediated by a supraoptimal dose of specific mono-clonal IgE antibody. Four-day PCl-immune cells were treated in vitro with anti-B220 plus complement to provide isolated late-acting DTH T cells. These cells were transferred alone (groups A and B), or into mice that also received an optimal dose of IgE (1 μg/mouse), or a supraoptimal dose of IgE (100 μg/mouse). Separate groups of these recipients were injected subcutaneously with cimetidine (50 mg/kg, 5 and 1 h prior to challenge), or were untreated, and then were challenged on the ears with PCl. A negative control group was challenged similarly and subtracted.

lenged on the ears by injection of various doses of serotonin and were also painted topically on the ears with PCl [14]. We found that a dose of 1,500 ng of serotonin locally caused a large early ear swelling response without appreciable late 24-hour reactivity and that DTH initiation occurred when this dose of serotonin was combined with late-acting T cells. Dose-response experiments revealed that local injection of as little as 50 or even 5 ng serotonin were sufficient to mediate DTH initiation. As with IgE, these very low doses of serotonin did not cause any detectable macroscopic early reactivity [14]. In fact, these low doses of serotonin were shown previously to be below those that could cause any local increase in vascular permeability, as detected by the extravasation of radiolabeled albumin [6]. Thus, vascular permeability itself is not necessary for DTH initiation.

The exact mechanism of DTH initiation by serotonin remains to be determined. In the context of the present experiments, the following are possibilities: (1) serotonin acts purely to vasodilate, which somehow allows local recruitment of late-acting T cells; (2) serotonin induces surface expression of adhesion molecules on endothelial cells; allowing for greater interaction with circulating T cells; (3) serotonin does not actually have a required vascular locus of action in DTH initiation, but is needed solely as a costimulus for the activation of the locally recruited T cells via their serotonin-2 receptors [10].

Further experiments will be necessary to distinguish these possibilities. We consider the above experiments on initiation of DTH by IgE and by serotonin to be of great significance for the following reasons: (1) they confirm the concept of DTH initiation; (2) DTH initiation has been achieved by employing defined mAb rather than molecularly uncharacterized DTH-initiating factors; (3) they raise the possibility that IgE antibody itself can normally, under some circumstances, initiate DTH; (4) they show the positive required role of serotonin and the negative role of histamine in murine DTH, and suggest a crucial mast cell involvement in these responses since *only* mast cells release *both* serotonin and histamine.

Are Mast Cells the Only Source of Serotonin in DTH?

The experiments above lead again to the question of whether mast cells are involved in DTH. Certainly the involvement in DTH of IgE, and IgG1 [14], the only immunoglobulin isotypes for which there are functional Fc receptors on mast cells, as well as the local release of *both* serotonin [5–11] and histamine [58], is highly suggestive of a role for mast cells. However, numerous experiments in two different strains of mast cell-deficient mice have not clearly demonstrated a crucial role of mast cells in DTH. The experiments of most investigators have shown that DTH is fully intact in both W/W^v and Sl/Sl^d mast cell-deficient mice [59–62]. However, our laboratory [63] and another [64] have shown that there is a partial deficiency of DTH in some experiments employing mast cell-deficient mice. Even in these experiments, DTH is not absent in mast cell-deficient mice, but is only reduced. However, and very importantly, even in the reduced DTH that is sometimes found in some experiments employing mast cell-deficient mice, this DTH is serotonin-dependent and thus inhibited by serotonin antagonists [63, 65]. Therefore, the crucial question that emerges is: If mast cells are involved in DTH, how do mast cell-deficient mice elicit DTH that is either fully intact or only partially defective?

This leads to the following question: What are the potential sources of local serotonin for DTH initiation in mast cell-deficient mice? The possibilities are as follows:

(1) *Platelets*: Platelets are well known to store large amounts of serotonin. In fact the function of platelet serotonin is unknown. We have found that beige mice, which have a defect in storing serotonin in platelets because of defective granules, have only 3% of blood serotonin compared to normal mice [Ratzlaff et al., in preparation]. Furthermore, staining ear and skin of normal mice with monoclonal antiserotonin antibodies revealed that there are only two common sites of serotonin localized in the skin: namely mast cell granules and intravascular platelets.

(2) *Blood basophils*: Until recently it was held that mice did not have basophils. However, it is now clear that a very small number of basophils exist in the bone marrow and blood of mice [66, 67], but there are no known basophils in the tissues of normal animals that could account for early serotonin release following antigen challenge to elicit DTH. In fact, although a basophil-rich variety of DTH known as cutaneous basophil hypersensitivity (CBH) is common in many responses of humans and guinea pigs, basophil infiltrates have not been found in various analogous forms of DTH in mice that have been studied by techniques that should be able to reveal basophils. However, one instance described recently, i.e. immune responses and rejection of ticks in the skin of mast cell-deficient W/W^v mice [68], has suggested that blood basophils can be recruited into the extravascular spaces, at least in this instance.

(3) *Mast cell precursors* are a third possible source of serotonin in mast cell-deficient mouse skin. It is known that mast cells arise from bone marrow-derived precursors that arrive in the skin as nondescript mononuclear cells which differentiate over 3–4 months into mature, granule-containing mast cells. Differentiation is regulated by local fibroblasts releasing stem cell factor, the ligand of the C-kit tyrosine kinase receptor on mast cell precursors [69, 70]. Thus, it is possible that mast cell precursors without granules exist in the skin and can synthesize and store serotonin in the cytosol. Serotonin in 'mast cells' would therefore not show up by stains for metachromatic granules, or even with mAb to serotonin. Indeed, experiments of Kitamura and co-workers in W/W^v mast cell-deficient mice demonstrating the absence of histamine-containing nonmast cells in the skin [71], and the induction of histamine-producing cells without basophil mast cell granules by topical

phorbol ester applied to the skin of mice [73], suggest that an immature mast cell subset or precursor could also produce serotonin that might be available in cytosolic vesicles for release in DTH initiation.

Employing a sensitive quantitative assay for serotonin, we found that mast cell-deficient mice, which have approximately 1% or less mast cells compared to normal mice, had not 1% but about 8% of the skin tissue serotonin compared to normal +/+ counterparts [Ratzlaff et al., in preparation]. Thus there is much more skin serotonin than can be accounted for by mast cells. This might be due to serotonin in mast cell precursors, but is likely due to serotonin in intravascular platelets.

(4) The fourth and final possibility is that the 1% or so mast cells that remain in the skin of a mast cell-deficient mouse might be sufficient for initiation of DTH. Thus, DTH-initiating factor, or IgE antibodies, could activate this small number of mast cells for a very low level release of serotonin that might not be associated with a macroscopically undetectable early response, but still could influence the local microenvironment to provide DTH initiation, as we have found in normal mice with very low doses of IgE, or local injection of very low doses of serotonin [14].

We have recently performed preliminary experiments suggesting that platelets may be an important source of serotonin for DTH initiation in mast cell-deficient mice. These experiments were made possible by the use of a polyclonal absorbed monospecific rabbit anti-mouse platelet antibody that specifically depletes mouse platelets by more than 97% for as long as 36 h after injection. For these experiments, mast cell-deficient W/W^v mice and their +/+ counterparts were treated with either absorbed antiplatelet antibody, or similarly absorbed normal rabbit serum, 5 h prior to eliciting DTH by PCl ear challenge. It was found that platelet depletion in mast cell-deficient mice resulted in a very significant diminution of DTH while there was only a mild diminution of DTH in +/+ mice that were treated with antiplatelet antibody [Geba et al., in preparation].

We have interpreted these data to indicate that platelets are an important source of serotonin for DTH initiation in mast cell-deficient mice, and may even play a role in DTH initiation of normal mice. The exact mechanism of platelet activation remains to be determined. There are at least two possibilities. Firstly, DTH-initiating factors or IgE antibodies may sensitize platelets via factor-binding sites, or via Fc-ε receptors such that at the site of local antigen challenge there is factor- or antibody-dependent platelet release of serotonin. Alternatively, it is possible that the factors and IgE *only* sensitize mast cells, and that the mast cells release not only serotonin but mediators that lead to release of serotonin by platelets in local vessels. This might be due to the action of serotonin, or other factors released by the mast cells, on the endothelial cells, thus rendering them more sticky for intravascular platelets which might then aggregate and release serotonin nonspecifically. Further experiments will be necessary to distinguish between these possibilities.

DTH Initiation Is Important in Other Systems

In the experiments that have been reviewed above, the concept of DTH initiation has been presented entirely in the context of murine contact hypersensitivity to PCl or OX. Several other laboratories [58, 73–76] have confirmed in contact sensitivity an early, serotonin-dependent, ear swelling initiating phase, featuring mast cell activation and vascular permeability. However, DTH initiation is not confined to these responses alone. Other systems in which DTH initiation has been identified are as follows:

(1) In C57Bl/6 (H-2^b) background mice that are immunized by low doses of intravenous sheep erythrocytes (SRBC), specific DTH is elicited 4 days later by SRBC challenge in the footpads. Experiments in immunized nude mice [36], and in immunized normal mice treated with serotonin antagonists [5–8], suggest that DTH initiation occurs in this system.

(2) In PPD-tuberculin DTH induced in mice by subcutaneous injection of PPD-pulsed macrophages, it has been shown that these responses include an early 2-hour reactivity that is induced within 1 day of immunization, and a late 24-hour reactivity that is detectable by 4 days. In contrast, PPD-pulsed splenic dendritic cells seem to induce no responses. However, when mice that are immunized by subcutaneous injection of PPD-pulsed dendritic cells, are immunized further with PPD-pulsed macrophages *1 day before* footpad challenge, then the early reactivity induced by the PPD-pulsed macrophages enables initiation of late 24-hour DTH [19]. The PPD-pulsed dendritic cells thus seem to *preferentially* induce late-acting DTH-effector T cells that *require* DTH initiation induced by the PPD-pulsed macrophages 1 day previously [19].

(3) In DTH to protein antigens, such as keyhole limpet hemocyanin, a Thy-1$^+$, CD5$^+$, CD3$^-$, CD4$^-$, CD8$^-$, B220$^-$ DTH 'helper cell' induces the CD4$^+$, CD3$^+$, DTH-effector Th-1 cell, *and* mediates an early 2-hour footpad response [77]. The DTH-helper cell resembles the DTH-initiating cell, but is B220$^-$, appears to be antigen non-specific, and despite its mediation of 2-hour

footpad swelling, has not yet been tested for whether it is *required* for the elicitation of 24-hour DTH.

(4) In mice recovering from a primary infection with the intestinal protozoa *Eimeria*, DTH footpad responses to oocyst antigen challenge are biphasic, with an early 2-hour and late 24- to 48-hour component. The ability to elicit *both* components is due to Thy-1[+], sIg[-], CD4[+] T cells [78]. This correlates with the ability to transfer primary immune resistance to the parasites with CD4[+] T cells [79], and raises the possibility that there are also CD4[+] DTH-initiating cells.

DTH Initiation and Tumor Immunity

A role for DTH initiation is indicated in several tumor systems in mice. Ultraviolet light (UV)-induced sarcomas in mice are known to each have their own tumor-specific antigens that probably are variations of MHC molecules. It has been found that mice immunized with a given UV-induced tumor develop the ability to elicit DTH responses to specific tumor cells. These responses are accompanied by an early 2-hour component which, like the late 24-hour component, is specific for the particular antigenic type of immunizing UV tumor [20]. Furthermore, when IL-2-dependent T cell lines specific for a given UV tumor cell were derived in vitro, it was found that these lines were not able to systemically transfer immune resistance to the tumors. In contrast, if the lines, which consisted of a combination of CD8[+] T cells (probably antitumor cytotoxic cells) and CD4[+] T cells (probably DTH-effector cells) was combined in transfer with 1-day immune cells corresponding to the same antigen-specific UV tumor, then recipients could mount immune resistance to the given tumor [20]. Therefore, antigen-specific DTH-initiating cells play a role in DTH to specific UV-induced tumors, and importantly, have a role in recruiting effector T cells of various types to mediate immune resistance to the tumors. We postulate that DTH initiation in tumor immunity, that may involve mast cell release of serotonin [18], not only recruits CD4[+] Th-1 T cells [80], but is also important in local recruitment of CD8[+] T cells, and possibly NK cells, as well as γδ T cells [81–84]. Thus, some forms of cellular immunity that are MHC class I-restricted might be called CD8[+] T cell DTH [85–91].

A further implication of DTH-initiating cells in immune responses and resistance to tumor cells is indicated by studies showing that an antigen-specific factor, which is induced *early* after immunization of mice with syngeneic or allogeneic tumor cells [92], can 'arm' macrophages for anti-tumor cell cytotoxicity. This specific macrophage-arming factor (SMAF) has many biological properties that are similar to PCl-F; the DTH-initiating factor for PCl contact sensitivity [92–96]. Furthermore, Vandebriel and co-workers have shown recently that SMAF activity and MHC allotumor antigen-specific early 2-hour footpad swelling were induced by immunization of nude athymic mice. Also, the phenotype of the SMAF-producing and early footpad swelling-producing cells (Thy-1[+], CD5[+], CD4[-], CD8[-]) is similar to DTH-initiating cells in contact sensitivity. However, a major difference is that the cells mediating the early component of MHC allotumor antigen-specific footpad swelling is reported to be CD3[+] [Vandebriel et al., submitted 1991]. This finding has led these workers to postulate that $\gamma\delta$ TCR-bearing T cells are involved. It is thus of great interest to await further studies by these investigators employing mAb that can specifically deplete $\gamma\delta$ versus $\alpha\beta$ T cells.

DTH Initiation and Mucosal Immune Responses to the Intestinal Nematode, Trichinella spiralis

Involvement of DTH-initiating cells in gastrointestinal inflammation is suggested in the model of *Trichinella spiralis* in mice, which features intestinal mucosal infiltrates of eosinophils and mast cells. It has been shown that *T. spiralis*-infested mice have early 2-hour cutaneous reactivity within 1–2 days following trichinella infestation. These early cutaneous responses are transferable with a Thy-1[+] CD5[+], CD8[-] cell population, and lymphoid cells harvested from *T. spiralis*-infested mice release a factor in vitro that can be concentrated on and eluted from *T. spiralis* antigen columns, and has biological properties that are analogous to the DTH-initiating factor PCl-F [21]. In this model, the intestinal inflammatory infiltrate of eosinophils and mucosal mast cells *is preceded* by antigen-specific local vascular permeability and recruitment of nonantigen-specific activated T lymphoblasts into the intestine [65]. All of these changes are serotonin-dependent and mast cell-dependent, and are inhibited by suppression of production of DTH-initiating factors [97]. We have concluded the following from these studies:

(1) DTH-initiating cells in the lymphoid organs make *T. spiralis* antigen-specific DTH-initiating factors that sensitize serotonin-containing cells in the gastrointestinal tract early after *T. spiralis* infestation for release of serotonin, following contact with *T. spiralis* antigens.

(2) Released serotonin serves a local DTH-initiating role that leads to extravascular recruitment of activated T lymphoblasts, some of which are CD4[+] T cells with a Th-2 lymphokine profile. Among these are a small number of *T. spiralis*-specific sensitized Th-2 cells that interact with *T. spiralis* antigen peptides complexed with class II MHC on local APC. Thus Th-2 cells are then activated to locally recruit circulating precursors of mucosal mast cells and then differentiate these precursors into mature mucosal mast cells, via release of IL-3 and IL-4, and possibly IL-10.

(3) Also, these recruited and antigen-specifically activated Th-2 cells release IL-5 and GM-CSF that lead to recruitment and activation of eosinophils.

These formulations account for the mast cell and eosinophil infiltrates that characterize these responses. Interestingly, local accumulation of IgE[+] and IgA[+] plasma cells is *not* serotonin-dependent, and may be due to the inherent mucosal recirculatory pattern of B cells that produce these isotypes.

Thus, we hypothesize that DTH-initiating cells release non-IgE factors that are important in the early phases of intestinal inflammation to Trichinella. Later these may be replaced by more professional IgE antibodies. It is noteworthy that ablation of these DTH-initiating responses in primary *T. spiralis* infestation does not interfere with primary immune resistance and expulsion of worms [97]. However, it remains a distinct possibility that secondary rapid expulsion is dependent on such factors, or IgE antibodies, acting in concert with recruitment to the epithelium of MHC class II-restricted CD4[+] T cells, as has been demonstrated recently in the *T. spiralis* model in rats [98, 99].

Conclusions about DTH Initiation in Mice

In summary, we have shown that DTH initiation in several systems in mice is due to primitive antigen-specific cells that make antigen-specific, antigen-binding factors that have biological activity analogous to IgE, but have molecular characteristics that are different. The non-IgE factors have yet to be characterized molecularly. We have also shown that IgE itself can initiate DTH and that in either case, serotonin is the crucial final mediator of DTH initiation, whereas histamine is suppressive. Finally, mast cells *and* platelets are both potential sources of the serotonin that mediates DTH initiation.

These findings are summarized in figure 5, showing that DTH-initiating cells make DTH-initiating factors like PCl-F or OX-F in the contact sen-

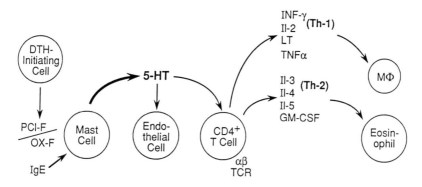

Fig. 5. DTH initiation can be mediated by non-IgE antigen-specific factors derived from primitive Thy-1⁺, CD5⁺, CD3⁻, sIg⁻, DTH-initiating cells or by IgE antibodies derived from B cells. DTH initiation leads to recruitment of CD4⁺ Th-1 cells (cutaneous DTH), or CD4⁺ Th-2 cells (mucosal DTH).

sitivity system, and that B cells may produce antigen-specific IgE antibody, either of which can sensitize mast cells for antigen-specific release of serotonin that is required early in DTH. Not included in figure 5 is the role that our preliminary results indicate for platelet release of serotonin.

Locally released serotonin has at least two loci of action in DTH. One locus is on endothelial cells, in some instances causing vascular permeability, but in other instances activating endothelial cells in other ways, such as expression of adhesion molecules. The other locus of action of serotonin is via serotonin-2 receptors on the recruited CD4⁺ cells. Here, serotonin seems to act as a costimulus that is required for successful production of cytokines by T cells that are simultaneously being activated via their αβ TCR. In cutaneous DTH it is apparent from studies with T cell clones, and from the lymphokine profile, that Th-1 cells are principally responsible for these responses, and that the important cytokines are interferon-γ [100–103] and TNFα [104], leading to recruitment and activation of monocytes, and in mice, neutrophils. On the other hand, in the gastrointestinal tract and perhaps in other mucosal surfaces, DTH initiation seems to lead to local recruitment of T cells with a Th-2 lymphokine profile [105–107] as indicated by the local accumulation of mucosal mast cells (IL-3, IL-4 and IL-10), and eosinophils (IL-5 and GM-CSF). Thus we suggest that *cutaneous* DTH in mice requires DTH initiation in which serotonin is the important final mediator leading to recruitment of T cells with a Th-1 cytokine profile, whereas *mucosal DTH*, as exemplified by the murine gastrointestinal tract responding to *T. spiralis*, also requires DTH initiation via local serotonin release, leading to recruitment of CD4+ T cells with a Th-2 lymphokine profile. It is possible that other mucosal sites also feature recruitment of CD4+ T cells with a Th-2 lymphokine profile. Certainly recent studies of T cell recruitment into the bronchial mucosa in patients with asthma [108], and into the conjunctiva in patients with vernal conjunctivitis [109], suggest a Th-2 cytokine profile, as these responses are associated with recruitment and activation of eosinophils, and local expression of IL-5 mRNA [110].

Does DTH Initiation Exist in Humans?

It is clear that T cells mediate DTH to proteins and contact sensitivity to haptens in humans and there is evidence that CD4+ Th-1 T cells [1, 111] and CD8+ [112] T cells are respectively important in these responses. However, an important unanswered question concerns whether DTH initiation exists in humans. Certainly there are no commonly described early *macroscopic* responses that accompany human DTH. However, our results in mice suggest that early macroscopic responses are not necessary [14]. It is noteworthy that early 2-hour vascular permeability and mast cell degranulation have been observed *microscopically* in human DTH responses [113–116]. Human mast cells do not contain and therefore cannot release serotonin, but we suggest that serotonin may not always be required, and that other mediators of vascular activation may be important in humans. Indeed, LTC_4 has been shown to be released during the early component of DTH in mice [117], and we have found that LTD_4 can also serve to initiate DTH in mice [Geba et al., in preparation]. Alternatively, activated rodent mast cells secrete an endoglycosidase that degrades heparan sulfate in subendothelial matrix [118], and such endoglycosidases have been suggested to be important in the extravasation of T cells in various forms of DTH in rats [119–121]. Thus, human cutaneous mast cells, or macrophages, could be a source of LTD_4 to mediate initiation of DTH, or released endoglycosidases could serve this function. Tumor necrosis factor (TNFα) and IL-1 are other important potential mediators of DTH initiation in humans. Within 2 h following activation of cutaneous mast cells in humans there is local endothelial cell expression of ELAM-1, an activation antigen important for endothelial adhesion of leukocytes [114, 115, 122]. Importantly, this induction of ELAM-1 is inhibited by treatment with cromolyn sodium [114], an inhibitor of mast cell secretion, and by antiserum to TNFα [115] and IL-1 [115, 122].

In addition, although serotonin is not contained in human mast cells, it is stored in human platelets which may therefore contribute to initiation of DTH. As suggested by our prior functional studies in mice [10], serotonin-2 receptors have recently been demonstrated on human T cells [123], and might serve to mediate a required co-stimulus in DTH. Histamine is the crucial vasoactive amine in humans. However, in humans, histamine probably does not serve a positive role, but as in mice, histamine may serve a negative function. Indeed, some studies indicate that treatment of humans with histamine-2 receptor antagonists tends to augment DTH responses [124–126].

A role for DTH initiation in human responses is suggested by novel new work showing that human T cells, secondarily activated in vitro with specific antigen, or within 1 day of priming in vitro, could elicit a 24-hour DTH response, when transferred to the footpads of mice [127]. This early DTH was due to T cells and was serotonin-dependent, and if the T cells were activated by antigen in vitro for 4 days then local adoptive transfer resulted in early 2-hour *and* late 24-hour components of DTH [127]. These results are strikingly similar to those obtained previously in immunized mice and suggest the presence *in humans* of antigen-specific DTH-initiating, and DTH-effector T cells, and furthermore suggest that adoptively transferred human DTH initiation is serotonin-dependent.

Finally, it is possible that physiologic levels of IgE could mediate DTH initiation in humans and thus lead to the elicitation of early responses that are often not macroscopically detectable, but still could act to influence the *micro*environment to allow local recruitment of CD4[+] DTH-effector T cells. Indeed, recent studies by Platts-Mills and co-workers [128] are very interesting in this regard. They studied DTH and IgE in subjects sensitive to *Trichophyton*, a common dermatophyte. Employing human IgE antibody as a probe to purify allergens of *Trichophyton*, these investigators identified Tri t-I as a major allergen in human IgE responses to *Trichophyton*. However, skin testing humans with this allergen, compared to raw *Trichophyton* extract, separated people into four groups. Firstly, there were those without any responses. Secondly, there were those with *only* early immediate wheal and flare responses. These individuals had specific IgE antibodies to Tri t-I, that were detectable by a sensitive radioimmunoassay. Interestingly, there were two other groups of patients that either had no early responses and just 'pure' DTH to *Trichophyton* antigen, *or* had *dual* reactivity with an immediate wheal and flare response, *followed by* late DTH responses. In *neither* of these latter two groups with DTH to *Trichophyton* was there any detectable IgE antibody to Tri t-I, nor to raw *Trichophyton* extract [128]. These results suggest that another antigen in *trichophyton* induces DTH-effector T cells that may require an early DTH-initiating component, that is mediated by either very low undetectable levels of IgE antibody, or by a non-IgE factor that is specific for this antigen. Further study will be necessary to distinguish between these interesting possibilities.

DTH Initiation and Autoimmunity

Experimental Autoimmune Encephalomyelitis (EAE)

Chronic inflammation due to the local recruitment into the tissues of CD4⁺ T cells is thought to be of crucial importance in several autoimmune diseases. EAE in rodents immunized with myelin basic protein (MBP) may be a model for multiple sclerosis. Experiments performed by Hinrichs et al. [129] in rat bone marrow MHC chimeras demonstrated that the endothelium did not play a role as an MHC-restricting element in the T cell transfer of EAE. This raised the question of how T cells were able to transfer EAE to a recipient in which the endothelium was allogeneic to the transferred cells. A postulated explanation was that T cells might activate mast cells to release serotonin, thus permitting recruitment across the endothelium. In support of these ideas were other studies demonstrating that clinical signs of EAE could be prevented in animals treated with drugs that altered responses to vasoactive amines [130–135]. In newer studies, Dietsch and Hinrichs [136] showed that T cells harvested from immunized rats and activated in vitro with antigen (or mitogen) transferred both EAE *and* DTH responses elicited by MBP. Importantly, the DTH consisted of an early 1-hour skin swelling and then subsequently a 24-hour swelling. The suggestion of an antigen-specific initiating phase of DTH to MBP led to experiments treating recipients with drugs that antagonized serotonin or altered serotonin storage, or inhibited mast cell release of mediators. The results of these experiments pointed to an early serotonin and mast cell-dependent initiating phase of EAE, and thus suggest that DTH-initiating cells play an important role in this model of autoimmune disease of the nervous system [136].

Adjuvant Arthritis

In an analogous model of rheumatoid arthritis in rats injected with mycobacterial adjuvant (called adjuvant arthritis), Trentham and co-workers

[137, 138] have demonstrated that synovitis can be elicited by an antigen (collagen)-specific factor derived from collagen-specific T cell lines and purified on collagen antigen affinity columns. Although the mechanism by which this factor induces a synovitis remains to be determined, electron microscopic studies showed that partial degranulation of the majority of local mast cells accompanied the early phases of synovial infiltration with inflammatory cells. Thus, the adjuvant arthritis model of rheumatoid arthritis may depend on a mast cell-activating, antigen-specific, DTH-initiating factor.

Experimental Autoimmune Uveoretinitis

In the model of experimental autoimmune uveoretinitis produced in rats immunized with bovine retinal antigen, susceptibility to disease correlates with choroidal mast cell numbers [139]. The fact that these mast cells undergo degranulation *prior* to the onset of uveitis [140] suggests that a DTH-initiating mechanism may play a role in the pathogenesis of this autoimmune disease.

Immune Diabetes mellitus

It has been established recently that the homing of antigen-specific T cells to the islets of Langerhans is important in autoimmune diabetes mellitus in the NOD mouse. Local recruitment of CD4$^+$ T cells and CD8$^+$ T cells has been demonstrated, and the homing of islet-destructive CD8$^+$ T cells [141] has been shown to be dependent on the CD4 T cells [142]. Recruitment of these effector T cell subsets to the islets may depend on prior activation of the local vessels as is suggested by vasopermeability in the islets. Furthermore, in the model of low dose streptozotocin diabetes in mice, it has been demonstrated that treatment with drugs that can block serotonin-dependent vascular permeability suppressed immune-mediated β cell destruction [143].

DTH Initiation in Immune Inflammation of the Lungs:
A New Mechanism in Asthma

Clinical Observations

Recent studies by Kay and co-workers [108] have established a potentially important role for T cells in immune inflammation of clinical asthma. Recruitment of activated CD4$^+$ T cells into asthmatic bronchi is associated with inflammation that features an infiltrate of activated eosinophils. This suggests that recruitment and activation of Th-2 T cells may be important in these responses. Furthermore, in atopic subjects, cells expressing mRNA of a Th-2 profile were found in allergen-induced cutaneous late phase reactions (LPR) containing activated CD4[+] T cells and eosinophils [144]. Furthermore, expression of IL-5 mRNA in mononuclear cells infiltrating mucosal bronchial biopsies from patients with asthma, correlated with the presence of activated (CD25[+]) CD4[+] T cells, and activated eosinophils [110].

An important question about these findings concerns the mechanisms for recruitment of Th-2 T cells and eosinophils and their relationship to airway narrowing and hyperreactivity in asthma, since infiltration and activation of eosinophils for release of mediators is thought to be of paramount importance in the pathogenesis of asthma. There are at least two possibilities. The first is that inhaled allergen triggers specific IgE antibodies to activate local mast cells (and/or macrophages) [145, 146] to release leukotrienes or cytokines [147–149] that attract and activate the T cells, *and* eosinophils. The second possibility is that allergen-specific T cells among the recruited T cells are themselves activated by Ag/MHC class II complexes on the local APC to in turn release cytokines that recruit and activate the eosinophils. The former possibility would ascribe the infiltrates to an IgE-dependent LPR, while the latter possibility, according to the data reviewed herein, likens the response to IgE-dependent DTH initiation that is required for recruitment of Th-2 T cells in a mucosal type of DTH. In fact, the differences may be semantic; both possibilities may coexist. Although it is probable in an actively sensitized individual, such as a patient with asthma, that IgE alone can mediate LPR, it is difficult to distinguish between these possibilities because passive IgE transfer versus adoptive T cell transfers to naive and challenged human recipients is not possible.

Experimental Observations in a Mouse Model of Asthma

A model of DTH and asthma in the lungs of mice suggests that non-IgE DTH-initiating factors, analogous to those produced by antigen-specific Thy-1[+] cells in contact sensitivity, may also be important in these responses. In this model, mice are contact-sensitized on the skin with PCl and challenged in the lung by intranasal administration of the analogous, water-soluble, reactive compound trinitrobenzene sulfonic acid (picryl sulfonic acid). In this model, lung DTH has been found to be T cell-dependent, serotonin-dependent, and in part, mast cell-dependent [150, 151]. A role for antigen-specific DTH-initiating cells and their produced antigen-specific

DTH-initiating factor, was suggested by suppression of lung DTH in mice rendered unable to produce DTH-initiating factors [150]. Further experiments suggest that airway hyperreactivity occurs in this model and is dependent on the DTH-initiating factors and the DTH-effector T cells [151].

Intranasal challenge with picryl sulfonic acid, to mice contact sensitized 1 week previously with PCl, induces an accumulation of inflammatory cells around airways and blood vessels. A DTH-like role for mast cells in these responses was suggested by reduction of lung DTH inflammation in mast cell-deficient W/W^v mice, but mast cells may not be required since another mast cell-deficient strain (Sl/Sl^d) showed only mild inhibition of inflammation [150]. Increased vascular permeability in the lungs was noted only 2 h after antigen challenge and treatment of these mice with serotonin receptor antagonists prevented the occurrence of DTH-like reactions in the lung [150, 152]. From these studies it was concluded that release of serotonin in an early DTH-initiating phase in the lung provided local microenvironmental changes, that included local vascular permeability, which facilitated the local recruitment and possible activation of DTH-effector T cells to subsequently attract infiltrates of inflammatory leukocytes.

A role for T cells in these responses was suggested by a lack of lung DTH in athymic nude mice, and by the adoptive transfer of lung DTH with Thy-1$^+$ lymphoid cells [151]. Importantly, in addition to DTH-like lung inflammation, mice with this model were shown to have increased pulmonary resistance in vivo [151]. Furthermore, isolated tracheas studied in vitro had hyperreactivity to carbachol within 2 h following challenge [151]. Tracheal hyperreactivity peaked 48 h after challenge, the time of maximum DTH infiltration, and lasted for at least 3 weeks. This local airway hyperreactivity was shown to be T cell-dependent since nude mice showed no hyperreactivity, and airway hyperreactivity was observed following adoptive transfer of sensitized T cells, and subsequent challenge in the lung [151].

These data strongly suggest that a T cell-dependent cascade of steps that begin with DTH-initiation is responsible for induction of airway hyperreactivity in this model of asthma. Furthermore, additional studies showed that mast cell-deficient mice had less severe airway hyperreactivity, and that depletion of transferring lymphoid cells with a mAb known to specifically deplete DTH-initiating cells abolished the ability to elicit airway hyperreactivity in recipients [152]. It was concluded that DTH-initiating cells had produced mast cell-sensitizing, DTH-initiating factors that, following local challenge, led to release of mediators, such as serotonin, that increased vascular permeability, which was important in the induction of airway inflammation, and importantly, airway hyperreactivity. Thus, airway hyper-reactivity detected in vitro and the change in lung resistance measured in vivo were induced by early DTH-initiating steps. In addition, transfer to naive recipients of PCl-F, the DTH-initiating factor, followed by local lung challenge, resulted in hyperreactivity of the trachea within 30 min, and increased vascular permeability and increased pulmonary resistance within 2 h after challenge [152]. From these data it was concluded that airway hyperreactivity and increased pulmonary resistance were induced by early DTH-initiating steps that depended on antigen-specific, mast cell-sensitizing, DTH-initiating factors.

More recent pharmacological studies have pointed to the importance of a DTH-initiating mediator *other* than serotonin in the pathogenesis of airway hyperreactivity. In these experiments it was shown that pretreatment with serotonin receptor-2 antagonists, such as ketanserin or methysergide, led to inhibition of lung DTH histological reactions, and diminished vascular permeability, but there was no effect on in vitro tracheal hyperreactivity to carbachol, and increased pulmonary resistance in vivo [153]. Thus, it was concluded that the presence of increased vascular permeability and inflammatory cell infiltrates were not necessary for the development of airway resistance and tracheal hyperreactivity. Therefore, the vascular permeability and inflammatory infiltrates appeared to be serotonin-dependent, but a serotonin-independent mechanism, that was mediated via the lung-acting DTH-initiating *factor*, was responsible for airway hyperresponsiveness and increased pulmonary resistance. It is also possible that DTH-initiating factors could activate lung macrophages to release cytokines, such as TNFα or IL-1 [154] to mediate or influence airway hyperreactivity.

One hypothesis to explain these interesting findings is that antigen-specific DTH-initiating factors activate mast cells for release of serotonin that is important in vascular permeability and the recruitment of inflammatory cells, but that other mast cell mediators, such as cytokines, are more important in inducing airway resistance and tracheal hyperreactivity in this model of asthma. Consistent with this formulation are recent findings demonstrating that nedocromil, an antiallergic and antiasthmatic drug known to inhibit mast cell release of mediators, and dexamethasone, a corticosteroid known to inhibit release of cytokines, are potent inhibitors of airway hyperresponsiveness in this model, whereas a platelet-activating factor antagonist and a cyclo-oxygenase inhibitor had no effect [155].

Taken together, the human data suggest that IgE-dependent late phase recruitment of Th-2 T cells is important in asthma, and the mouse lung model suggests that non-IgE DTH-initiating factors may play an important role. To synthesize the findings of these systems we hypothesize that IgE plays an important role in DTH initiation in human asthma. Furthermore, the potential importance of IgE as a DTH-initiating factor in asthma may explain some characteristic clinical aspects of this illness. It is well known that allergic asthmatic patients have IgE antibodies to typical environmental inhalant allergens such as pollens, house dust mites, and animal danders. Under these circumstances mild asthma may be largely dependent on IgE-mediated immediate inflammatory responses, and on LPR featuring recruitment of Th-2 T cells specific for the allergens in question. However, a common occurrence clinically is for such patients to be in a stable condition, controlled by bronchodilators and topical corticosteroids, but then, when infected locally by viruses or bacteria, to have a profound exacerbation of their asthma; often leading to Emergency Room visits and hospitalizations, necessitating systemic treatment with corticosteroids.

Aspects of DTH initiation by IgE offer an explanation for this scenario. It has been shown in the murine cutaneous model of DTH that the early-acting IgE antibodies and late-acting DTH T cells need not be of the same antigen specificity if the skin is challenged with *both* antigens (fig. 6). Thus, in the lung of patients with asthma, environmental allergens may provide a background of IgE-dependent DTH-initiating mechanisms that ordinarily lead to the recruitment into the lungs of a few CD4$^+$ (or CD8$^+$) T cells specific for antigens derived from infectious agents, but these cells are not activated in the absence of specific infection. However, during a viral or bacterial infection these *few* T cells that are recruited by DTH-initiating mechanisms due to allergen-specific IgE, meet local APC that have processed viral or bacterial antigen peptides and present them on surface MHC molecules. Thus, with infection, the few recruited T cells specific for peptide antigens derived from the infectious agent are activated to produce more profound local inflammation, edema and mediator release that is associated with a great exacerbation of clinical asthma. This requires systemic treatment with corticosteroids, which have a dominant locus of action on inhibiting production of cytokines by the recruited, antigen-activated T cells. If this hypothesis is correct, then treatment with cyclosporin, or possibly the more potent FK506, which are immunosuppressants that inhibit T cell transcription and release of cytokines, may be beneficial in severe asthma, and could have a high therapeutic benefit if given topically into the lungs.

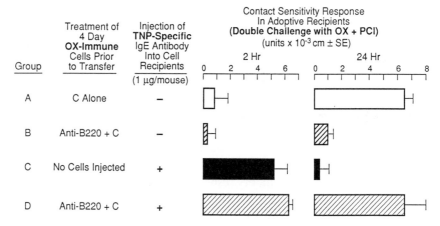

Fig. 6. IgE that mediates initiation of DTH can be of a different Ag specificity from late DTH T cells. Four-day OX-immune cells were treated with complement alone (group A) to serve as positive controls, or with anti-B220 plus complement to provide isolated late-acting DTH T cells. These OX-immune late DTH T cells were transferred alone into mice (group B), or to mice that had received 1 µg anti-TNP IgE antibody 24 h previously. These recipients, and a group that received IgE alone (group C) were challenged topically on the ears with *a mixture* of 0.8% PCl + 0.8% OX. A negative control group was challenged similarly with a mixture of PCl + OX.

Conclusions

The essence of DTH is recruitment of antigen-specific lymphokine-producing effector T cells into the extravascular tissues at local sites of antigen challenge. Recruitment depends on DTH initiation that mediates activation of mast cells and probably platelets to release the vasoactive amine serotonin. A crucial aspect of DTH initiation is that mast cells release serotonin in an antigen-specific manner due to *prior* sensitization by antigen-specific non-IgE DTH-initiating factors, *or* via IgE antibodies. The antigen-specific non-IgE DTH-initiating factors are produced in the lymphoid organs within 1 day of immunization by primitive, thymic-independent cells with an unusual phenotype for antigen-specific cells (Thy-1+, CD5+, CD4-, CD8-, CD3-, TCRαβ-, TCRγδ-, sIg-, CD45RA+, Mac1+, IL-2R- and IL-3R+). The genes encoding the antigen-specific non-IgE DTH-initiating factors are *neither* TCR nor Ig genes, and may belong to another family of rearranging genes.

Very low doses of specific IgE antibodies (1–10 ng/mouse) can also mediate DTH initiation, while high doses of IgE (10–100 μg/mouse) *inhibit* DTH initiation via release of histamine in large amounts, along with serotonin. Histamine acts via histamine-2 receptors, probably to inhibit transcription and thus production of lymphokines by the recruited DTH-effector T cells.

DTH initiation has been demonstrated in contact sensitivity, PPD-tuberculin hypersensitivity, tumor-specific immune resistance, and in intestinal inflammation responses to the nematode parasite *T. spiralis*. Whereas *cutaneous DTH* leads to local recruitment of Th-1 CD4+ T cells, *mucosal DTH* as exemplified by intestinal responses to *T. spiralis*, and perhaps bronchial responses in asthma, leads to recruitment of Th-2 CD4+ T cells. In responses to tumors, DTH initiation leads to recruitment of CD4+ and CD8+ effector T cells, and also probably NK cells, and possibly γδ T cells as well. DTH-initiating recruitment of CD4+ T cells is important in models of autoimmunity involving neuroantigens, such as EAE and uveoretinitis. DTH initiation may also play a role in other autoimmune models, such as adjuvant arthritis and immune diabetes mellitus.

DTH initiation may occur in humans. Recent clinical evidence suggests that recruitment of Th-2 T cells is important in bronchial asthma. Experimental evidence in a mouse model suggests that lung inflammation and airway narrowing and hyperresponsiveness are dependent on DTH-initiating mechanisms. It is postulated that some exacerbations of clinical allergic asthma by bronchial viral or bacterial infections are due to allergen-specific IgE-dependent recruitment of DTH-effector T cells with specificity for viral or bacterial antigen peptides that are presented by local APC. If this is true, then *local topical* treatment with drugs such as cyclosporin or FK506, which inhibit generation of cytokines, may have a great therapeutic benefit in this disease.

References

1 Platt JL, Grant BW, Eddy AA, Michael AF: Immune cell populations in cutaneous delayed-type hypersensitivity. J Exp Med 1983;158:1227.
2 Cher DJ, Mossman TR: Two types of murine helper T cell clone. II. Delayed-type hypersensitivity is mediated by T_H1 clones. J Immunol 1987;138:3688.
3 Herzog WR, Ferreri NR, Ptak W, Askenase PW: The antigen-specific DTH-initiating Thy-1+ cell is double negative (CD4⁻, CD8⁻) and CD3 negative, and expresses IL-3 receptors, but no IL-2 receptors. J Immunol 1989;143:3125–3133.

4 Marchal G, Seman M, Milon G, Truffa-Bachi P, Zilberfarb V: Local adoptive transfer of skin delayed-type hypersensitivity initiated by a single T lymphocyte. J Immunol 1982;129:954.

5 Gershon RK, Askenase PW, Gershon M: Requirement for vasoactive amines in the production of the skin reactions of delayed-type hypersensitivity. J Exp Med 1975;42:732–747.

6 Schwartz A, Askenase PW, Gershon RK: The effect of locally injected vasoactive amines on the elicitation of delayed-type hypersensitivity. J Immunol 1977;118: 159–165.

7 Askenase PW, Bursztajn S, Gershon MD, Gershon RK: T cell dependent mast cell degranulation and release of serotonin in murine delayed-type hypersensitivity. J Exp Med 1980;152:1358–1374.

8 Askenase PW, Metzler CM, Gershon RK: Localization of leukocytes in DTH reaction sites and lymph nodes: dependence on vasoactive amines. Immunology 1982;47:239–246.

9 Van Loveren H, Kraeuter-Kops S, Askenase PW: Different mechanisms of release of vasoactive amines by mast cells occur in T cell-dependent compared to IgE-dependent cutaneous hypersensitivity responses. Eur J Immunol 1984;14:40–47.

10 Ameisen JC, Meade R, Askenase PW: A new interpretation of the involvement of serotonin in delayed-type hypersensitivity: Serotonin-2 receptor antagonists inhibit contact sensitivity by an effect on T cells. J Immunol 1989;142:3171–3179.

11 Kraeuter Kops S, Van Loveren H, Rosenstein RW, Ptak W, Askenase PW: Mast cell activation and vascular alterations in immediate hypersensitivity-like reactions induced by a T cell-derived antigen binding factor. Lab Invest 1984;50:421–434.

12 Ptak W, Herzog W-R, Askenase PW: Delayed-type hypersensitivity initiation by early-acting cells that are antigen mismatched or MHC incompatible with late-acting, delayed-type hypersensitivity effector T cells. J Immunol 1991;46:469–475.

13 Herzog W-R, Ferreri NR, Ramabhadran R, Millet I, Askenase PW: An antigen-specific DTH-initiating cell clone: Functional phenotypical and partial molecular characterization. J Immunol 1990;144:3667–3676.

14 Ptak W, Geba GP, Askenase PW: Initiation of delayed-type hypersensitivity by low doses of monoclonal IgE antibody. Mediation by serotonin and inhibition by histamine. J Immunol 1991;146:3929–3936.

15 Van Loveren H, Meade R, Askenase PW: An early component of delayed-type hypersensitivity mediated by T cells and mast cells. J Exp Med 1983;157:1604–1617.

16 Van Loveren H, Askenase PW: Delayed-type hypersensitivity is mediated by a sequence of two different T cell activities. J Immunol 1984;133:2397–2401.

17 Van Loveren H, Kato K, Meade R, Green DR, Horowitz M, Ptak W, Askenase PW: Characterization of two different Lyl⁺ T cell populations that mediate delayed-type hypersensitivity. J Immunol 1984;33:2402–2411.

18 Van Loveren H, Den Otter W, Meade R, Terheggen PMA, Askenase PW: A role for mast cells and the vasoactive amine serotonin in T cell dependent immunity to tumors. J Immunol 1985;134:1292–1299.

19 Mukherjee S, Katz DR, Rook GAW: Differing role of dendritic cells and macrophages in the induction of delayed-type hypersensitivity responses to PPD. Immunology 1986;59:229.

20 Trial J: Cooperation between early-acting delayed-type hypersensitivity T cells and cultured effector cells in tumor rejection. Cancer Res 1988;48:5922–5926.

21 Parmentier HK, Dijkstra JW, Wissink A, Ruitenberg EJ, Askenase PW, Van Loveren H: Identification and partial characterization of a T cell-derived antigen-binding factor from mice infected with the intestinal helminth, *Trichinella spiralis*. Int Arch Allergy Appl Immunol 1989;90:237–247.

22 Ptak W, Askenase PW, Rosenstein RW, Gershon RK: Transfer of an antigen specific immediate hypersensitivity-like reaction with an antigen binding factor produced by T cells. Proc Natl Acad Sci USA 1982;79:1969–1973.

23 Askenase PW, Rosenstein RW, Ptak W: T cells produce an antigen binding factor with in vivo activity analogous to IgE antibody. J Exp Med 1983;157:862–873.

24 Askenase PW, Van Loveren H, Rosenstein RW, Ptak W: Immunologic specificity of antigen-binding T cell-derived factors that transfer mast cell dependent, immediate hypersensitivity-like reactions. Monogr Allergy. Basel, Karger, 1983, vol 18, pp 249–255.

25 Kraeuter Kops S, Ratzlaff RE, Meade R, Iverson GM, Askenase PW: Interaction of antigen-specific T cell factors with unique 'receptors' on the surface of mast cells: demonstration in vitro by an indirect rosetting technique. J Immunol 1986;136:4515–4524.

26 Van Loveren H, Ratzlaff RE, Kato K, Meade R, Ferguesson T, Iverson M, Janeway CA, Askenase PW: Immune serum from mice contact sensitized with picryl chloride contains an antigen-specific T cell factor that transfers immediate cutaneous reactivity. Eur J Immunol 1986;16:1203–1208.

27 Meade R, Van Loveren H, Parmentier H, Iverson GM, Askenase PW: The antigen-binding T cell factor PCl-F sensitized mast cells for in vitro release of serotonin: Comparison with monoclonal IgE antibody. J Immunol 1988;141:2704–2713.

28 Theoharides TC, Bondy PK, Tsakalos ND, Askenase PW: Differential release of serotonin and histamine from mast cells. Nature 1982;297:229–231.

29 Tamir H, Theoharides TC, Gershon MD, Askenase PW: Serotonin storage pools in basophil leukemia and mast cells: characterization of two types of serotonin binding protein and radioautographic analysis of the intracellular distribution of ^{3}H serotonin. J Cell Biol 1982;93:638–647.

30 Theoharides TC, Kraeuter Kops S, Bondy PK, Askenase PW: Differential release of serotonin without comparable histamine can occur under diverse conditions in the rat mast cell. Biochem Pharmacol 1985;34:1389–1398.

31 Kraeuter Kops S, Theoharides T, Kashgarian MG, Askenase PW: Ultrastructural characteristics of peritoneal mast cells undergoing differential release of serotonin without histamine, and without degranulation. Cell Tissue Res 1990;262:415–424.

32 Carraway RE, Cochrane DE, Granier C, et al: Parallel secretion of endogenous 5-hydroxytryptamine and histamine from mast cells stimulated by vasoactive peptides and compound 48/80. Br J Immunol 1984;81:227–229.

33 Vliagoftis H, Dimitriadou V, Theoharides TC: Progesterone triggers selective mast cell secretion of 5-hydroxytryptamine. Int Arch Allergy Appl Immunol 1990;93:113–119.

34 Moorhead JW: Antigen receptors on murine T lymphocytes in contact sensitivity. J Exp Med 1981;154:1811–1826.

35 Garssen J, van Loveren H, Kato K, Askenase PW: In vitro desensitization of immunized cells with hapten-amino acid vs. hapten-protein conjugates suggests that there are different antigen receptors on the two Thy-1+ cells that mediate the early and late components of murine contact sensitivity (submitted 1991).

36 Herzog W, Meade R, Pettinicchi A, Ptak W, Askenase PW: Nude mice produce a
    T cell-derived antigen-binding factor that mediates the early component of delayed-
    type hypersensitivity. J Immunol 1989;142:1803–1812.
37 Lieber MR, Jesse JE, Lewis S, Bosma GS, Rosenberg J, Mizuuchi K, Bosma MJ,
    Gellert M: The defect in murine severe combined immune deficiency: joining of
    signal sequences but not coding segment in V(D)J recombination. Cell 1988;55:7.
38 Malynn G, Blackwell T, Fulop G, Rathbun G, Furley A, Ferrier P, Heinke L, Phillips
    R, Yancoupoulos G, Alt F: The SCID defect affects the final step of the immunoglob-
    ulin VDJ recombinase mechanism. Cell 1988;54:453.
39 Schatz DG, Oettinger MA, Baltimore D: The V(D)J recombination activating gene,
    RAG-1. Cell 1989;59:1035.
40 Herzog WR, Ptak W, Askenase PW: Suppression and contrasuppression in athymic
    nude mice: Nude mice produce the antigen-binding component of an antigen-
    specific T suppressor factor that inhibits the late 24-hour component of DTH, but do
    not generate isotype-like suppression or contrasuppression of the early initiating
    phase of DTH. Cell Immunol 1990;127:130–145.
41 Millet I, Ferreri NR, Ramabhadran R, Askenase PW: IL-3 dependence of a Thy-1^{lo},
    B220$^+$, Mac-1$^+$, IL-3 receptor positive antigen-specific DTH-initiating clone (sub-
    mitted 1991).
42 Askenase PW, Hayden BJ, Gershon RK: Augmentation of delayed-type hypersensiti-
    vity by doses of cytoxan which do not affect antibody responses. J Exp Med 1975;
    141:697–702.
43 Sy M-S, Miller SD, Claman HN: Immune suppression with supraoptimal doses of
    antigen in contact sensitivity. I. Demonstration of suppressor cells and their sen-
    sitivity to cyclophosphamide. J Immunol 1977;119:240–244.
44 Maquire HC Jr, Faris L, Weidanz W: Cyclophosphamide intensifies the acquisition
    of allergic contact dermatitis in mice rendered B-cell deficient by heterologous anti-
    IgM antisera. Immunology 1979;37:367–372.
45 Chiorazzi N, Fox D, Katz DH: Hapten-specific IgE antibody responses in mice. VI.
    Selective enhancement of IgE antibody production by low doses of x-irradiation and
    by cyclophosphamide. J Immunol 1976;117:1629.
46 Graziano F, Haley C, Gunderson L, Askenase PW: IgE antibody production
    of guinea pigs treated with cyclophosphamide. J Immunol 1981;127:1067–
    1070.
47 Dwyer JM, Rosenbaum JT, Lewis S: The effect of anti-μ suppression of IgM and IgG
    on the production of IgE. J Exp Med 1976;143:781.
48 Yaoita Y, Kumagai Y, Okumura K, Honjo T: Expression of lymphocyte surface IgE
    does not require switch recombination. Nature 1982;297:697.
49 Chan MA, Benedict SH, Dosch H-M, Hui MF, Stein LD: Expression of IgE from a
    nonrearranged ε locus in cloned B-lymphoblastoid cells that also express IgM.
    J Immunol 1990;144:3563–3568.
50 Mercolino TJ, Arnold LW, Hawkins LA, Haughton G: Normal mouse peritoneum
    contains a large population of Ly-1$^+$ (CD5$^+$) B cells that recognize phosphatidylcho-
    line. Relationship to cells that secrete hemolytic antibody specific for autologous
    erythrocytes. J Exp Med 1988;168:687–698.
51 Snapper CM, Hooley JJ, Barbieri S, Finkelman FD: Murine B cells expressing
    Thy-1 after in vivo immunization selectively secrete IgE. J Immunol 1990;144:
    2940–2945.

52 Askenase PW, Schwartz A, Siegel J, Gershon RK: The role of histamine in the regulation of cell mediated immunity. Int Arch Allergy Appl Immunol 1981;66 (suppl 1):225.

53 Tasaka K, Kurokawa K, Nakayama Y, Nakimoto M: Effect of histamine on delayed-type hypersensitivity in mice. Immunopharmacology 1986;12:69.

54 Rocklin RE: Modulation of cellular immune responses in vivo and in vitro by histamine receptor-bearing lymphocytes. J Clin Invest 1976;57:1051.

55 Dohlsten M, Sjögren HO, Carlsson R: Histamine inhibits interferon-γ production via suppression of interleukin-2 synthesis. Cell Immunol 1986;101:103.

56 Dohlsten M, Sjögren HO, Carlsson R: Histamine acts directly on human T cells to inhibit interleukin-2 and interferon-γ production. Cell Immunol 1987;109:115.

57 Arad G, Nussinovich R, Kaempfer R: Dual control of human interleukin-2 and interferon-γ gene expressed by histamine: activation and suppression (submitted 1991).

58 Kerdel FA, Belsito DV, Scotto-Chinnici R, Soter NA: Mast cell participation during the elicitation of murine allergic contact hypersensitivity. J Invest Dermatol 1987; 88:686.

59 Thomas WR, Schrader JW: Delayed hypersensitivity in mast cell deficient mice. J Immunol 1983;130:2565.

60 Galli SJ, Hammel I: Unequivocal delayed hypersensitivity in mast cell-deficient and beige mice. Science 1984;226:710.

61 Ha T-Y, Reed ND, Crowle PK: Immune response potential of mast cell-deficient W/W^v mice. Int Arch Allergy Appl Immunol 1986;80:85–94.

62 Mekori YA, Chang JCC, Wershil BK, Galli S: Studies of the role of mast cells in contact sensitivity responses. Cell Immunol 1987;109:39.

63 Askenase PW, Van Loveren H, Kraeuter-Kops S, Ron Y, Meade R, Theoharides TC, Norlund JJ, Scovern H, Gershon MD, Ptak W: Defective elicitation of delayed-type hypersensitivity in W/W^v andf Sl/Sl^d mast cell deficient mice. J Immunol 1983;131: 2687–2694.

64 Miyachi Y, Imamura S, Tokura Y, Takigawa M: Mechanisms of contact photosensitivity in mice. VII. Diminished elicitation by reserpine and defective expression in mast cell-deficient mice. J Invest Dermatol 1986;87:38.

65 Parmentier HK, Garssen J, Vos JG, Askenase PW, Van Loveren H: Involvement of T-cell-derived antigen binding factors and serotonin in intestinal vasopermeability and accumulation of lymphoblasts during the early phase of a Trichinella spiralis infection (submitted 1991).

66 Urbina C, Ortiz C, Hurtado I: A new look at basophils in mice. Int Arch Allergy Appl Immunol 1981;66:158–160.

67 Dvorak AM, Nabel G, Pyne K, Cantor H, Dvorak HF, Galli SJ: Ultrastructural identification of the mouse basophil. Blood 1982;59:1279–1285.

68 Steeves EBT, Allen JR: Basophils in skin reaction of mast cell-deficient mice infected with Dermacentor variabilis. Int J Parasitol 1991.

69 Copeland NG, Gilbert DJ, Cho BC, Donovan PJ, Jenkins NA, Cosman D, Anderson D, Lyman SD, Williams DE: Mast cell growth factor maps near the steel locus on mouse chromosome 10 and is deleted in a number of steel alleles. Cell 1990;63:175–183.

70 Zsebo KM, Wypych J, McNiece IK, Lu HS, Smith KA, Karkare SB, Sachdev RK, Yuschenkoff VN, Birkett NC, Williams LR, Satyagal VN, Tung W, Bosselman RA,

Mendiaz EA, Langley KE: Identification, purification, and biological characterization of hematopoietic stem cell factor from buffalo rat liver-conditioned medium. Cell 1990;63:195–201.

71 Yamatodani A, Maeyama K, Watanabe T, Wada H, Kitamura Y: Tissue distribution of histamine in a mutant mouse deficient in mast cells. Clear evidence for the presence of non-mast-cell histamine. Biochem Pharmacol 1982;31:305–309.

72 Taguchi Y, Tsuyama K, Watanabe T, Wada H, Kitamura Y: Increase in histidine decarboxylase activity in skin of genetically mast-cell-deficient W/Wᵛ mice after application of phorbol-12-myristate-13-acetate: Evidence for the presence of hista mine-producing cells without basophilic granules. Proc Natl Acad Sci USA 1982;79: 6837–6841.

73 Ishii N, Ikezawa Z, Nagai R, Okuda K: Ir gene control of murine contact hypersensitivity. Jpn J Dermatol 1980;90:1337–1341.

74 Ray MC, Tharp MD, Sullivan TJ, Tigelaar RE: Contact hypersensitivity reactions to dinitrofluorobenzene mediated by monoclonal IgE anti-DNP antibodies. J Immunol 1983;131:1096–1102.

75 Mekori YA, Galli SJ: [¹²⁵I]fibrin deposition occurs at both early and late intervals of IgE-dependent or contact sensitivity reactions elicited in mouse skin. Mast cell-dependent augmentation of fibrin deposition at early intervals in combined IgE-dependent and contact sensitivity reactions. J Immunol 1990;145: 3719.

76 Lavaud P, Rodrique F, Carré C, Touvay C, Mencia-Huerta J-M, Braquet P: Pharmacologic modulation of picryl chloride-induced contact dermatitis in the mouse. J Invest Dermatol 1991;97:101.

77 Matsushima GK, Stohlman SA: Distinct subsets of accessory cells activate Thy-1⁺ triple negative (CD3⁻, CD4⁻, CD8⁻) cells and Th-1 delayed-type hypersensitivity effector T cells. J Immunol 1991;146:3322–3331.

78 Shi Y, Mahrt JL, Mogil RJ: Kinetics of murine delayed-type hypersensitivity response to *Eimeria falciformis* (Apicomplexa: Eimeriidae). Infect Immun 1989;57: 146–151.

79 Rose ME, Josey HS, Hesketh P, Grencis RK, Wakelin D: Mediation of immunity to *Eimeria veriformis* in mice by L3T4⁺ T cells. Infect Immun 1988;56:1760–1765.

80 Nagarkatti M, Clary SR, Nagarkatti PS: Characterization of tumor-infiltrating CD4⁺ T cells as Th1 cells based on lymphokine secretion and functional properties. J Immunol 1990;144:4898.

81 Maghazachi AA, Goldfard RH, Herberman RB: Influence of T cells on the expression of lymphokine-activated killer cell activity and in vivo tissue distribution. J Immunol 1988;141:4039–4046.

82 Karpati RM, Banks SM, Malissen B, Rosenberg SA, Sheard MA, Weber JS, Hodes RJ: Phenotypic characterization of murine tumor-infiltrating T lymphocytes. J Immunol 1991;146:2043–2051.

83 Berd D, Murphy G, Maquire HC Jr, Mastrangelo MJ: Immunization with haptenized, autologous tumor cells induces inflammation of human melanoma metastases. Cancer Res 1991;51:2731–2734.

84 Zangemeister-Wittke U, Kyewski B, Schirrmacher V: Recruitment and activation of tumor-specific immune T cells in situ. CD8⁺ cells predominate the secondary response in sponge matrices and exert both delayed-type hypersensitivity-like and cytotoxic T lymphocyte activity. J Immunol 1989;143:379–385.

85 Sunday M, Benacerraf B, Dorf ME: Hapten-specific T cell responses to 4-hydroxy-3-nitrophenyl acetyl. VI. Evidence for different T cell receptors in cells that mediate H-21-restricted and H-2D-restricted cutaneous sensitivity responses. J Exp Med 1980;152:1554.

86 Minami M, Okuda K, Sunday ME, Dorf ME: H-2K-, H21- and H-2D-restricted hybridoma contact sensitivity effector cells. Nature 1982;297:231–233.

87 Lin Y-L, Askonas BA: Biological properties of an influenza A virus-specific killer T cell clone. Inhibition of virus replication in vivo and induction of delayed-type hypersensitivity reactions. J Exp Med 1981;154:225–234.

88 Kelly CJ, Korngold R, Mann R, Clayman M, Haverty T, Neilson EG: Spontaneous interstitial nephritis in kdkd mice. II. Characterization of a tubular antigen-specific, H-2K-restricted Lyt-2$^+$ effector T cell that mediates destructive tubulointerstitial injury. J Immunol 1986;136:526.

89 Kaufman SH: CD8$^+$ T lymphocytes in intracellular microbial infections. Immunol Today 1988;9:168.

90 Doherty PC, Allan JE, Lynch F, Ceredig R: Dissection of an inflammatory process induced by CD8$^+$ T cells. Immunol Today 1990;11:55.

91 Allan W, Tabi Z, Cleary A, Doherty PC: Cellular events in the lymph node and lung of mice with influenza. J Immunol 1990;144:3980–3986.

92 Dullens HF, DeWeger RA, Van Der Maas M, Den Besten PJ, Vandebriel RJ, Den Otter W: Production of specific macrophage-arming factor precedes cytotoxic T lymphocyte activity in vivo during tumor rejection. Cancer Immunol Immunother 1989;30:28–33.

93 Los G, De Weger RA, Moberts RMP, Van Loveren H, Sakkers RJ, Den Otter W: The absence of delayed-type hypersensitivity reactivity in a syngeneic murine tumor system. Immunology 1987;62:89.

94 Vandebriel RJ, De Weger RA, Los G, Van Loveren H, Wiegers GJ, Weernink PAO, Den Otter W: Two specific T cell factors that initiate immune responses in murine allograft systems: A comparison of biologic functions. J Immunol 1989;143:66–73.

95 De Weger RA, Vandebriel RJ, Slager H, Mans D, Van Loveren H, Wilbrink B, Dullens HFJ, Den Otter W: Initial immunochemical characterization of specific macrophage-arming factor. Cancer Immunol Immunother 1989;30:21–27.

96 De Groot JW, De Weger RA, Vandebriel RJ, Den Otter W: Differences in the induction of macrophage cytotoxicity by the specific T lymphocyte factor, specific macrophage-arming factor, and the lymphokine, macrophage activating factor. Immunobiology 1989;179:131–144.

97 Parmentier HK, Dijkstra W, Wissink A, Ruitenberg RJ, Askenase PW, Van Loveren H: Antigen-specific T cell factors induce isotype-like suppression of mast cell and eosinophil-rich T cell-dependent inflammation in the intestine of mice infected with *Trichinella spiralis*. Int Arch Allergy Appl Immunol 1989;90:144–154.

98 Wang CH, Korenaga M, Sacuto FR, Ahmad A, Bell RG: Intraintestinal migration to the epithelium of protective, dividing, anti-*Trichinella spiralis* CD4$^+$ OX22$^-$ cells requires MHC class II compatibility. J Immunol 1990;145:1021–1028.

99 Ahmad A, Wang CH, Bell RG: A role for IgE in intestinal immunity. Expression of rapid expulsion of *Trichinella spiralis* in rats transfused with IgE and thoracic duct lymphocytes. J Immunol 1991;146:3563.

100 Kaplan G, Luster AD, Hancock G, Cohn ZA: The expression of a γ-interferon-induced protein (IP-10) in delayed immune responses in human skin. J Exp Med 1987;166:1098.

101 Issekutz TB, Stoltz JM, Meide PVD: Lymphocyte recruitment in delayed-type hypersensitivity. The role of IFN-γ. J Immunol 1988;140:2989.
102 Fong AT, Mossmann TR: The role of IFN-γ in delayed-type hypersensitivity mediated by Th1 clones. J Immunol 1989;143:2887.
103 Gessner A, Drjupin R, Löhler J, Lother H, Lehmann-Grube F: IFN-γ production in tissues of mice during acute infection with lymphocytic choriomeningitis virus. J Immunol 1990;144:3160.
104 Piguet PF, Grau GE, Hauser C, Vassalli P: Tumor necrosis factor is a critical mediator in hapten-induced irritant and contact hypersensitivity reactions. J Exp Med 1991;173:673.
105 Mossmann RR, Cherwinski H, Bond MW: Two types of murine helper T cell clones. I. Definition according to profiles of lymphokine activities and secreted proteins. J Immunol 1986;136:2348.
106 Bottomly K: A functional dichotomy in CD4+ T lymphocytes. Immunol Today 1988;9:268.
107 Coffman RL, Seymour BWP, Hudak S: Antibody to interleukin-5 inhibits helminth-induced eosinophilia in mice. Science 1989;245:308–310.
108 Assawi M, Bradley B, Jeffrey PK, Frew AJ, Wardlaw AJ, Knowles G, Assoufi B, Collins JV, Durham S, Kay AB: Activated T lymphocytes and eosinophils in bronchial biopsies in stable atopic asthma. Am Rev Respir Dis 1990;142:1407–1413.
109 Maggi E, Biswas P, Prete GD, Parronchi P, Macchia D, Simonelli C, Emmi L, DeCarli M, Tiri A, Ricci M, Romagnani S: Accumulation of Th-2-like helper T cells in the conjunctiva of patients with vernal conjunctivitis. J Immunol 1991;146:1169–1174.
110 Hamid Q, Azzawi M, Ying S, Moqbel R, Wardlaw AJ, Corrigan CJ, Bradley B, Durham SR, Collins JV, Jeffrey PK, Quint DJ, Kay AB: Expression of mRNA for interleukin-5 in mucosal bronchial biopsies from asthma. J Clin Invest 1991;87:1541–1546.
111 Tsicopoulos A, Hamid Q, Varney V, Ying S: Th1-type cells (mRNA IFN-γ+, IL-2+, IL-4−, IL-5−) in classical delayed-type (tuberculin) hypersensitivity responses in human skin (submitted 1991).
112 Kalish RS, Johnson KL: Enrichment and function of urushiol (poison ivy)-specific T lymphocytes in lesions of allergic contact dermatitis to urushiol. J Immunol 1990;145:3706–3713.
113 Lewis RE, Buchshaum M, Whitaker MS, Murphy GF: Intercellular adhesion molecule expression in the evolving human cutaneous delayed hypersensitivity reaction. J Invest Dermatol 1989;93:672–677.
114 Klein LM, Lavker RM, Matis WL, Murphy GF: Degranulation of human mast cells induces an endothelial antigen central to leukocyte adhesion. Proc Natl Acad Sci USA 1989;86:8972.
115 Walsh LJ, Trinchieri G, Waldorf HA, Whitaker D, Murphy GF: Human dermal mast cells contain and release tumor necrosis factor α which induces endothelial leukocytes adhesion molecule-1. Proc Natl Acad Sci USA 1991;88:4220.
116 Dvorak AM, Mihm MC Jr, Dvorak HF: Morphology of delayed-type hypersensitivity reaction in man. II. Ultrastructural alternatives affecting the microvasculature and the tissue mast cells. Lab Invest 1976;34:179–191.
117 Meurer R, Opas EE, Humes JL: Effects of cyclooxygenase and lipoxygenase inhibitors on inflammation associated with oxazolone-induced delayed hypersensitivity. Biochem Pharmacol 1988;37:3511–3514.

118 Bashkin P, Razin E, Eldor A, Vlodavsky I: Degranulating mast cells secrete an endoglycosidase that degrades heparan sulfate in subendothelial extracellular matrix. Blood 1990;75:2204–2212.

119 Naparstek Y, Cohen IR, Fuks Z, Vlodavsky I: Activated T lymphocytes produce a matrix-degrading heparan sulphate endoglycosidase. Nature 1984;310:241–244.

120 Lider O, Baharav E, Mekori YA, et al: Suppression of experimental autoimmune diseases and prolongation of allograft survival by treatment of animals with low doses of heparins. J Clin Invest 1989;83:752–756.

121 Lider O, Mekori YA, Miller T, Bar-Tana R, Vlodavsky I, Baharav E, Cohen IR, Naparstek Y: Inhibition of T lymphocyte heparanase by heparin prevents T cell migration and T cell-mediated immunity. Eur J Immunol 1990;20:493–499.

122 Leung DYM, Pober JS, Cotran RS: Expression of endothelial-leukocyte adhesion molecule-1 in elicited late phase allergic responses. J Clin Invest 1991;87: 1805.

123 Aune TM, Kelley KA, Ranges GE, Bombara MP: Serotonin-activated signal transduction via serotonin receptors on jurkat cells. J Immunol 1990;145:1826–1831.

124 Avella J, Binder H, Madsen JE, Askenase PW: Effect of histamine H2-receptor antagonists on delayed hypersensitivity. Lancet 1978;i:624.

125 Jorizzo JL, Sams WM, Jegasothy BU, Olansky AJ: Cimetidine as an immunomodulator. Chronic mucocutaneous candidiasis as a model. Ann Intern Med 1980;92: 192.

126 Brockmeyer NH, Kruzefelder E, Mertins L, Chalabi N, Kirch W, Scheiermann N, Goos M, Ohnhaus EE: Immunomodulatory properties of cimetidine in ARC patients. Clin Immunol Immunopathol 1988;48:50.

127 Trial JA: Adoptive transfer of early and late delayed-type hypersensitivity reactions mediated by human T cells. Reg Immunol 1989;2:14–21.

128 Deuell B, Arruda LK, Hayden ML, Chapman MD, Platts-Mills TAE: *Trichophyton tonsurans* allergen. I. Characterization of a protein that causes immediate but not delayed hypersensitivity. J Immunol 1991;147:96–101.

129 Hinrichs DJ, Wegmann KW, Dietsch GN: Transfer of experimental allergic encephalomyelitis to bone marrow chimeras. Endothelial cells are not a restricting element. J Exp Med 1987;166:1906–1911.

130 Linthicum DS, Frelinger JA: Acute autoimmune encephalomyelitis in mice. II. Susceptibility is controlled by the combination of H-2 and histamine sensitization genes. J Exp Med 1982;155:31–40.

131 Linthicum DS: Development of acute autoimmune encephalomyelitis in mice: factors regulating the effector phase of the disease. Immunobiology 1982;162:211–220.

132 Linthicum DS, Munoz JJ, Blasket A: Acute experimental allergic encephalomyelitis in mice. I. Adjuvant action of *Bordetella pertussis* is due to vasoactive amine sensitization and increased vascular permeability of the central nervous system. Cell Immunol 1982;73:299.

133 Goldmuntz EA, Brosnan CF, Norton WT: Prazosin treatment suppressed increased vascular permeability in both acute and passively transferred experimental autoimmune encephalomyelitis in the Lewis rat. J Immunol 1986;137:3444.

134 Brosnan CF, Sacks HJ, Goldschmidt RC, Goldmuntz EA, Norton WT: Prazosin treatment during the effector stage of disease suppresses experimental autoimmune encephalomyelitis in the Lewis rat. J Immunol 1986;137:3451.

135 Waxman FJ, Taguiam JM, Whitacre CC: Modification of the clinical and histo-pathologic expression of experimental allergic encephalomyelitis by the vasoactive amine antagonist cyproheptidine. Cell Immunol 1984;85:82–93.

136 Dietsch GN, Hinrichs DJ: The role of mast cells in the elicitation of experimental allergic encephalomyelitis. J Immunol 1989;142:1476–1481.

137 Helfgott SM, Kieval RI, Breedveld FC, Brahn E, Young CT, Dynesius-Trentham R, Trentham DE: Detection of arthritogenic factor in adjuvant arthritis. J Immunol 1988;140:1838–1843.

138 Caulfield JP, Hein A, Helfgott SM, Brahn E, Dynesius-Trentham RA, Trentham DE: Intraarticular injection of arthrogenic factor causes mast cell degranulation, inflam-mation, fat necrosis and synovial hyperplasia. Lab Invest 1988;59:82–95.

139 Mochizuki M, Kuwabara T, Chan C-C, Nussenblatt RB, Metcalfe DD, Gery I: An association between susceptibility to experimental autoimmune uveitis and choroi-dal mast cell numbers. J Immunol 1984;133:1699–1701.

140 de Kozak Y, Sainte-Laudy J, Benveniste J, Faure J-P: Evidence for immediate hypersensitivity phenomena in experimental autoimmune uveoretinitis. Eur J Im-munol 1981;11:612–617.

141 Nagata M, Yokono K, Hayakawa M, Kawase Y, Hatamori N, Ogawa W, Yonezawa K, Shii K, Baba S: Destruction of pancreatic islet cells by cytotoxic T lymphocytes in nonobese diabetic mice. J Immunol 1989;143:1155–1162.

142 Thivolet C, Bendelac A, Bedossa P, Back J-F, Carnaud C: CD8+ T cell homing to the pancreas in the non-obese diabetic mouse is CD4+ T cell-dependent. J Immunol 1991;146:85–88.

143 Schwab E, Burkart V, Freytag G, Kiesel U, Kolb H: Inhibition of immune-mediated low-dose streptozotocin diabetes by agents which reduce vascular permeability. Immunopharmacology 1986;12:17–21.

144 Frew AJ, Kay AB: The relationship between infiltrating CD4+ lymphocytes, acti-vated eosinophils, and the magnitude of the allergen-induced late phase cutaneous reaction in man. J Immunol 1988;141:4158–4164.

145 Rankin JA, Hitchcock M, Merrill WW, Huang SS, Brashler JR, Bach MK, Askenase PW: IgE immune complexes induce immediate and prolonged release of leukotriene C4 from rat alveolar macrophages. J Immunol 1984;132:1993–1999.

146 Borish L, Mascali JJ, Rosenwasser LJ: IgE-dependent cytokine production by human peripheral blood mononuclear phagocytes. J Immunol 1991;146:63–67.

147 Plaut M, Pierce JH, Watson CJ: Mast cell lines produce lymphokine in response to cross-linkage of FcεRI or to calcium ionophores. Nature 1989;339:64.

148 Wodnar-Filipowicz A, Heusser H, Moroni C: Production of the hemopoietic growth factors GM-CSF and interleukin-3 by mast cells in response to IgE receptor-mediated activation. Nature 1989;339:150.

149 Burd PR, Rogers HW, Gorden JR, et al: Interleukin-3-dependent mast cells stimu-lated with IgE and antigen express multiple cytokines. J Exp Med 1989;170:245.

150 Garssen J, Nijkamp FP, Wagenaar SS, Zwart A, Askenase PW, Van Loveren H: Regulation of delayed-type hypersensitivity-like responses in the mouse lung, deter-mined with histological procedures: serotonin, T cell suppressor-inducer factor and high antigen dose tolerance regulate the magnitude of T cell dependent inflammatory reactions. Immunology 1989;68:51–58.

151 Garssen J, Nijkamp FP, Van Der Vliet H, Van Loveren H: T-cell mediated induction of airway hyperreactivity in mice. Am Rev Respir Dis 1991;144:931–938.

152 Garssen J, Nijkamp FP, Van Vugt E, Van Der Vliet H, Van Loveren H: A possible role for initiating steps in cellular immunity in the induction of airway hyperreactivity (submitted 1991).

153 Garssen J, Van Loveren H, Van Der Vliet H, Bot H, Nijkamp FP: Delayed-type hypersensitivity-induced airway hyperresponsiveness and altered lung functions in mice are independent of increased vascular permeability and mononuclear cell infiltration (submitted 1991).

154 Ferreri NR, Millet I, Paliwal V, Herzog W, Solomon D, Ramabhadran R, Askenase PW: Induction of macrophage TNFα, IL-1, IL-6, and PGE$_2$ production by DTH-initiating factors. Cell Immunol 1991;137:389.

155 Garssen J, Van Loveren H, Van Der Vliet H, Nijkamp FP: Pharmacological modulation of T cell dependent airway hyperresponsiveness in mice (submitted 1991).

Philip W. Askenase, MD, Section of Allergy and Clinical Immunology,
Department of Medicine, Yale University School of Medicine,
333 Cedar Street, New Haven, CT 06510 (USA)

Subject Index